Fish Ecology, Evolution, and Exploitation

MONOGRAPHS IN POPULATION BIOLOGY

SIMON A. LEVIN AND HENRY S. HORN, SERIES EDITORS

Fish Ecology, Evolution, and Exploitation

A New Theoretical Synthesis

KEN H. ANDERSEN

PRINCETON UNIVERSITY PRESS
Princeton and Oxford

Copyright © 2019 by Princeton University Press
Published by Princeton University Press
41 William Street, Princeton, New Jersey 08540
6 Oxford Street, Woodstock, Oxfordshire OX20 1TR

press.princeton.edu

All Rights Reserved

Library of Congress Control Number 2018957496

Cloth ISBN 978-0-691-17655-0
Paperback ISBN 978-0-691-19295-6

British Library Cataloging-in-Publication Data is available

Editorial: Alison Kalett and Kristin Zodrow
Production Editorial: Sara Lerner
Text and Cover Design: Carmina Alvarez
Production: Erin Suydam
Publicity: Matthew Taylor and Julia Hall
Copyeditor: Jennifer Harris

This book has been composed in Times Roman

Printed on acid-free paper. ∞

Printed in the United States of America

10 9 8 7 6 5 4 3 2 1

Contents

PART III
TRAITS

Preface

Thanks to all the colleagues, collaborators, students, funders, and everybody else whose presence and support has made it possible for me to write this book. The theory is the product of my work during more than a decade at the National Institute for Aquatic Resources at the Technical University of Denmark. I am grateful for being part of the broad and inspiring scientific environment that the institute has fostered. None of my work would have been possible without support from my other close colleagues and collaborators: Robert Arlinghaus, Jan Beyer, Julia Blanchard, Keith Brander, Keith Farnsworth, Henrik Gislason, Simon Jennings, Thomas Kiørboe, Jake Rice, Uffe H. Thygesen, and Andy Visser, and my students: Martin Hartvig, Nis Sand Jacobsen, Alexandros Kokkalis, Karin Olsson, and Rob van Gemert. Funding to support the work has been provided through numerous EU projects: MEECE, MYFISH, IMAGE, FACTS and FishAce, and in particular by the Villum Foundation through its generous support of the Centre for Ocean Life.

This book was written between other duties over a two-year period. Particular thanks to Jan Beyer for getting me started and to Thomas Kiørboe for pushing me to finish it. Thanks to Daniel van Denderen, Christian Jørgensen, and Marc Mangel for comments, and to Rob van Gemert for patiently proofing all the text (remaining errors are solely my responsibility). Thanks also to Simon Levin and his inspiring group of students and postdocs for hosting me at Princeton University during spring 2016, where the first half of the book was written, and to the Danish Shellfish Center for hosting me during the final stages. Last but not least, thanks to my wife, Anna, for her patience and support.

Notation

TABLE 0.1. Notation.

Symbol	Description	Value	Reference
A	Growth coefficient	$5.35 \text{ g}^{1/4}/\text{yr}$	p. 51
a	Physiological mortality	0.42	Section 4.4
β	Preferred predator:prey mass ratio	408	p. 25
β_{PPMR}	Predator:prey mass ratio in stomach	708	Table 2.2
$B_c(w)$	Community biomass spectrum	# g/g	p. 87
γ	Coefficient for clearance rate	$\text{g}^{-q}\text{V}/\text{yr}$	p. 22, Table C.1
C	Consumption rate	g/yr	p. 24, Eq. 2.7
C_{max}	Maximum consumption rate	g/yr	p. 24
c	Constant in length-weight relation	0.01 g/cm^3	p. 19
ε_a	Assimilation efficiency	0.6	p. 50, eq. 3.27
ε_{egg}	Reproductive efficiency	0.22	p. 47
ε_R	Recruitment efficiency	0.03	p. 71
E_a	Available energy	g/yr	p. 167
E_e	Encountered food	g/yr	Eq. 10.2
f_0	Expected average feeding level	0.6	p. 31
f_c	Critical feeding level	0.2	p. 50
η_F	w_F/W_∞ for trawl	0.05	p. 86
Φ_a	Coefficient for available food	—	Eq. 2.7
Φ_p	Coefficient for mortality	—	Eq. 2.15
$\phi(w_p/w)$	Prey size preference	—	p. 25
$g(w)$	Growth rate	g/yr	Eq. 3.18, eq. 10.4
$G_{\text{rs}.\theta}$	Relative specific selection response	yr^{-1}	Eq. 6.10
h	Coefficient for maximum consumption	$22.3 \text{ g}^{1-n}/\text{yr}$	p. 24
h^2	Heritability	0.2	p. 106
κ_c	Coefficient of the community size spectrum	$\text{# g}^{\lambda-1}$	Eq. 2.21
κ_{res}	Carrying capacity of resource spectrum	$\text{# g}^{\lambda-1}$	Eq. 10.12
K	von Bertalanffy growth constant	yr^{-1}	p. 40
K_{Rmax}	Constant for maximum recruitment	0.25	p. 190

(Continued)

TABLE 0.1. (*continued*)

Symbol	Description	Value	Reference
k	Specific reproductive investment	yr^{-1}	Eq. 3.17
l	Length	cm	p. 40
L_∞	Asymptotic length	cm	p. 40
λ	Exponent of the abundance size spectrum	2.05	Eq. 2.21
μ_p	Predation mortality	yr^{-1}	Eq. 2.22, eq. 10.16
μ_F	Fishing mortality	yr^{-1}	p. 84
$N(w)$	Population number spectrum	#/g	p. 61
$N_c(w)$	Community number spectrum	#/g	Eq. 2.5, eq. 2.20
N_{res}	Resource number spectrum	#/g	
n	Exponent for maximum consumption	3/4	p. 24
$\psi_m(z)$	Maturation function	—	Eq. 3.15
$P_{w_1 \to w_2}$	Survival from w_1 to w_2	—	Box 4.3
q	Exponent for clearance rate	0.8	p. 22
R	Recruitment flux	# per time	p. 72, eq. 4.36
R_0	Lifetime reproductive output "eggs per recruit"	—	Eq. 4.39
R_{egg}	Individual reproductive output	g/yr/#	Eq. 3.19
R_{max}	Maximum recruitment	# per time	Eq. 4.36
R_p	Population reproductive output	# per time	Eq. 4.35
r_{max}	Population growth rate	per time	Section 7.1
σ	Width of prey size selection function	1	p. 25
σ_F	Width of gill net selectivity	1.5	Eq. 5.4
t_{mat}	Age at maturation	yr	Eq. 3.25
u	Sharpness of trawl selectivity	3	p. 85, eq. 5.3
$V(w)$	Clearance rate	V/yr	p. 22
w	Body wet weight	g	p. 18
w_0	Egg weight	1 mg	Fig. 8.2
w_F	Characteristic size of retainment of fishing gear	g	p. 85
w_m	Weight at maturation	g	p. 45, Fig. 3.4
w_R	Weight at recruitment	0.001 g	p. 71
W_∞	Asymptotic weight	g	Eq. 3.11
ξ	Constant for starvation mortality	1	p. 169
Y	Fisheries yield	g/yr	Eq. 5.7
Y_R	Fisheries yield per recruit	g/#/yr	p. 91

NOTE: # refers to numbers, yr to years. Measures of abundance and biomass can refer to either the total number or biomass in a system or to concentration. See p. 60.

Fish Ecology, Evolution, and Exploitation

CHAPTER ONE

Nothing as Practical as a Good Theory

This book presents a mathematical theory of fish stocks and fish communities. The theory describes the demography of fish stocks, the structure of fish communities, and the evolutionary ecology of fish. Throughout, the theory is applied to relevant problems in fisheries science: impact of fishing on demography, fisheries reference points, evolutionary impact assessments, stock recovery, ecosystem-based fisheries management, and so on, as well as to basic ecological and evolutionary questions: population growth rate, density dependence, offspring size, and the like. Before going into the details of the theory, some context is needed: Why do we need a new theory? Which problems should it address? How do we formulate such a theory?

Fish are the dominant marine organisms in the body size range from about 1 g to 100 kg. They inhabit all the worlds' oceans, from the sunlit surface waters to the darkest depths, and in freshwater they are able to find niches in even in the smallest lakes and rivers. Their exceptional high productivity makes them an important source of food and wealth for humans. The Food and Agriculture Organization (FAO, 2016) estimates that fisheries provide about 10 percent of global human consumption of protein at a value of about $100 billion/yr. Despite fish being highly productive and fecund, modern fisheries have been capable of overexploiting fish stocks since the advent of modern trawler technology in the mid-twentieth century. To maintain high yields, fisheries therefore have to be managed. Because fish are hidden from plain sight beneath the surface of the oceans, fisheries management relies on mathematical models to assess the impact of fishing on fish stocks and develop efficient fishing and management strategies. The theoretical background for such models was developed in the first half of the twentieth century on the basis of age-structured matrix models and condensed into the *Beverton and Holt framework* from 1949 (fig. 1.1). Today, most advice for fisheries management is supported by the Beverton and Holt framework; however, its age is showing, and it is coming under increased pressure.

Fisheries management faces several challenges. First, it struggles to implement the "Ecosystem Approach to Fisheries Management," laid down in the Reykjavik

FIGURE 1.1. Ray Beverton (with pipe) and Sidney Holt (front) at work in 1949. Photo by Michael Graham. *Source:* Ramster (1996) ICES J. Mar. Sci., 53:1–9.

declaration from 2001. The ecosystem approach mandates that current single-stock-oriented management is extended toward managing the entire ecosystem. The Beverton and Holt framework is geared toward managing single stocks, and new model tools are needed to deal with multispecies aspects. Second, management faces new questions: What are the long-term evolutionary consequences of the selection imposed by fishing? How should it deal with the large fraction of "data poor" fish stocks, particularly in the developing world, where no or little biological information exists? How should it handle the many ecosystems that are very species diverse, where fisheries are largely indiscriminate toward species, making management on a stock-by-stock basis impractical?

An obvious place to look for help and inspiration would be general ecology. However, because of the need to specialize, fisheries science has become isolated and disjoint from ecology. After Beverton and Holt published their framework, fisheries science branched away from general ecology and concentrated its efforts on operationalizing the framework to practical application for management. Fisheries science developed its own conferences, published much important research in the gray literature of conference proceedings or working group reports, and created its own specialized journals. In the meantime, ecology sprouted new branches, particularly in limnology (inland aquatic ecosystems), food-web ecology, structured populations, and evolutionary ecology, all of which could be relevant for fisheries science.

Among fish ecologists, the main action is in limnology. Pure marine fish ecologists are rare, as most are engaged with the fisheries practice. An exception is the study of coral reef fish, which are a special case not much treated here. The advantage of working in lakes—in particular, small ones—is that their ecosystems are easier to observe and understand because of their low diversity and low habitat complexity. Within theoretical population ecology, a notable development is physiological structured models (Metz and Diekmann, 1986), which generalize classic consumer-resource models to structured populations, and are particularly applicable to fish. These advances in understanding lake ecosystems have had next to no impact on fisheries science in the seas. There have been some attempts at convincing fisheries scientists to adopt insights and techniques from freshwater fish (Persson et al., 2014), but with little success; fisheries scientists seem not to appreciate the advanced insights and theories, as they cannot be easily operationalized in practical fisheries management. To reach out to fisheries science, limnologists must face the difficulties of working in the seas and the messy business of implementing fisheries management.

Other novel branches of ecology that would be relevant for fisheries science are food-web ecology and evolutionary ecology. The advent of computers made it possible to generalize simple competition or predator-prey models to entire food webs, and an entire discipline emerged to study such complex food-web models. The discipline homed in on questions of structure and stability to identify general patterns in the topology of food webs (who eats whom) and which types of structures make a food web stable. The discussion was largely about identifying general rules or statistical patterns, and there has been little attention to developing models of specific food webs of particular ecosystems. Further, the question of how food webs responds to perturbations, such as fishing, was never central. A notable exception is the EcoPath type of models, which has indeed been occupied with setting up food-web models of specific systems, and such models are also increasingly used in fisheries science. However, overall fisheries science and management have not been able to assimilate the developments in theoretical food-web ecology.

Fish have had a special place in the hearts of evolutionary ecologists, and evolutionary ecologists probably see fish in the broadest context. The idea of "life history invariants" was born through observations of fish (Beverton, 1992) and later generalized by Charnov et al. (2001). Central evolutionary problems in fish ecology are to understand the diversity of offspring size strategies, reproductive strategies, and the evolution of indeterminate growth. While evolutionary ecology has been central to understanding fish life histories, it has found little application in fisheries science.

Against the backdrop of the challenges to fisheries management and the increasing interest from classic ecology and evolutionary ecology in fish and fisheries, this book introduces the size- and trait-based approach as a modern, coherent, and unifying framework to model fish populations and fish communities. The theory is woven from strands taken from newer developments in ecology and fisheries science that will make it applicable broadly to fisheries and ecological problems. By catering to both fisheries scientists and ecologists, I hope to contribute to the long overdue unification of thinking in fish ecology and fisheries science. I will now describe the basic elements of the theory, starting with those elements coming from classic fisheries science—in particular, with regard to applications—and then moving on to size-based theory as developed in marine ecology, physiologically structured population models, and trait-based ecology.

Fisheries science and management is the most important application of the theory. In the context of fisheries science, the theory can be seen as a reformulation of the traditional single-stock Beverton and Holt framework from scratch. It is tempting to repair the Beverton and Holt framework and add some missing pieces to make it applicable to the ecosystem approach to fisheries management. That would be like constructing a car by welding two bikes together and adding an engine. Repairing Beverton and Holt would make it impossible to achieve the degree of theoretical rigor that I strive for. I believe, as does Kurt Lewin, who coined the quote in this chapter's title, that practical applications, like fisheries advice, are best given from a solid theoretical basic understanding. Starting over with a new theory entails throwing out classic concepts like the treasured von Bertalanffy growth equation with the ubiquitous K and L_∞ parameters, doing away with spreadsheet-friendly life tables, and scrapping the concepts of adult mortality, M and $M2$, to mention just a few. Instead of von Bertalanffy, I use physiology; instead of life tables, I use differential equations; and instead of the constant adult mortality, I use a size-based mortality. The absence of well-known concepts may make the theory appear inaccessible and overly complicated to one well-versed in the classics of fisheries science, such as described by Hilborn and Walters (1992) or Quinn and Deriso (1999). The reward is a theory that is consistently built upon a few fundamental assumptions, from which it deals with classic single-stock impact assessment, but also estimates evolutionary rates and makes ecosystem impact assessments.

Others have reformulated the Beverton and Holt framework. In a nineteenth-century castle housing the Danish Institute for Fisheries and Marine Research, K. P. Andersen[1] and Erik Ursin were toiling away in the 1970s. They wanted to bring the Beverton and Holt single-species framework into the multispecies

---

[1] No relation!

reality of real marine ecosystems. And they succeeded. Unfortunately, the theory was too complex, and it fizzled out. The equations themselves fill several pages (Andersen and Ursin, 1977). Not only that, but the numerical implementation of a complex model was a major undertaking at the time—it had to be coded on punch cards! Along the way, Andersen and Ursin introduced several important novel ideas: everything is based on a description of the physiology of individual fish, accountance of all mass flows—including primary-secondary production and recycling—and size-based selection of prey. Most of their work is forgotten because it was published in obscure journals—for example, Ursin (1979) in *Symposia of the Zoological Society of London*, or the now folded Danish journal *Dana*.

I combine Andersen and Ursin's ideas with size-based theory. The importance of body size as a central structuring component of ecology and evolution has been recognized for at least a century. I rely upon the scaling of metabolism with body size, referred to as Kleiber's law (Kleiber, 1932), and the rule that big fish eat smaller fish. Sheldon and co-workers showed how these two rules combine to explain body-size distributions (Sheldon et al., 1977), and the ideas were later used to develop the building blocks of dynamic models (Silvert and Platt, 1978). The *metabolic theory* (Brown et al., 2004) made similar predictions; however, I go further that the dimensional arguments in metabolic theory and I provide a stronger mechanistic foundation for some of the metabolic scaling rules—in particular, mortality. I also predict the size structure within populations, and not just within communities. Much of the work on size-based population demography builds on the pioneering efforts by Jan Beyer (1989). A surprising result is that some of the metabolic scaling rules actually do not apply as expected for fish population, despite the reliance on metabolic scaling on the level of individual organisms. This is important, as such rules are widely used formally or implicitly.

While fisheries science was largely content with developing the Beverton and Holt framework toward practical applications, ecologists continued their fundamental inquiry into the dynamics of fish populations—in particular, in limnology. A crucial juncture is the review by Werner and Gilliam (1984). Just as Beverton and Holt did, Werner and Gilliam stressed the importance of describing the entire life cycle of fish, and not just the adults. However, they also realized how the age-based Beverton and Holt theory was unable to describe the complicated interactions of competition and predation between different stages of fish populations. Interactions occur mainly because of differences in body size, not age, and these interactions lead to density-dependent bottlenecks. They then sketched a new theoretical framework based on body size instead of age. Lennart Persson and André de Roos bought Werner and Gilliam's vision about density-dependent bottlenecks and managed to surpass the formidable analytical challenges to develop applications of physiologically structured populations (De Roos and Persson, 2013). To create a theory directed toward fisheries applications, I focus on another aspect of Werner and Gilliam's vision—namely, the development of size-structured population dynamics. A similar development is integral projection models (Easterling et al., 2000), which are essentially discrete versions of the continuous time- and size-based demography that I develop here.

With regard to life-history theory, there is a fascinating analogy between the life histories of plants and fish. Both groups share three notable characteristics: they (mostly) make very small offspring; they (mostly) do not have parental care; and they continue to grow after maturation. There are other reasons for looking for inspiration in plant ecology. Plant ecologists have developed trait-based approaches that cut through the complexity of dealing with the myriads of species making up a plant community. Instead of describing each species separately, they rather characterize the distribution of the main traits of species in a community. This approach turns out to be very powerful when dealing with entire fish communities. Trait-based approaches are controversial—how can you throw away species, when species are at the core of fisheries management and biology? After all, Darwin wrote about the origin of species, not about the origin of traits. This is a valid concern. I use the idea of traits to generalize across all species; however, much of the theory on the population level can equally well be applied to particular species.

I found the inspiration to develop the trait-based framework for fish in the work of John Pope and co-workers (2006). They related all species-specific parameters to the average maximum size that individuals in each species can obtain. That crucial insight made the asymptotic (maximum) size into a master trait. Characterizing differences between species just by their asymptotic size opens the door to making broad statements about all fish species just by sweeping over asymptotic sizes. Of course, using only one trait is a gross simplification, and the trait-based approach can be generalized by including more traits than just the asymptotic size. Nevertheless, the central idea is to characterize species by just a few fundamental traits, so the introduction of additional traits must be done with care. The trait-based approach is particularly important for developing a dynamic theory of the entire fish community because it circumvents the complexity of having to deal with a tangled food web of many interacting species. It is also the secret ingredient that makes the theory particularly relevant to data-poor situations, because no matter how little we know about a specific stock, we have a good idea of the maximum size of landed individuals. Last, the trait-based approach is a powerful tool to obtain insights that have broad validity. However, one should not be dogmatic about it—real ecosystems actually do consist of species, and practical fisheries management must care about specific stocks. Therefore, the single-species model I present can equally well be applied to specific stocks, and I show how the trait-based community model can be formulated as a species-based food-web model.

1.1 WHAT CHARACTERIZES A GOOD THEORY?

A good theory can be likened to a game of cards. A game of cards is defined by a few simple rules that can be explained quickly over a coffee table. If the rules are well chosen, they define a complex and entertaining game. Similarly, a theory is based upon a few fundamental axioms. The axioms must be generally accepted and have a solid empirical foundation or relations to other theory. A good theory makes nontrivial predictions of both qualitative and quantitative nature. For example, a good theory about fish stocks not only predicts that some level of exploitation extracts the maximum yield from the stock, but it also predicts the actual level of fishing mortality that maximizes yield.

Fish ecology is challenged by the difficulty of carrying out controlled experiments. Let's compare with an idealized version of physics. In physics, theory goes hand in hand with experiments: experiments makes discoveries, theory proposes an explanation and possibly additional hypotheses, and experimentalists go back to check the explanation and the new hypotheses. Things are not quite that straightforward in ecology because experiments are less accessible. Physicists can create idealized experimental conditions where most confounding effects are eliminated or accounted for. In ecology, such conditions may be obtained while describing the physiology of individual organisms—for example, the functional response may be measured through the feeding of organisms at different food concentrations, or the swimming speed and respiration may be measuring in a flow chamber. For experiments with entire communities, however, clean conditions are out of reach. And that is not even considering the issue of time scales—the time scales of change of ecological communities are longer than the longest-lived individuals in the community, typically on the order of decades. Because of these fundamental difficulties, experiments are rare and only possible in a few cases and at great effort, such as in lakes (for example, Carpenter et al., 1987; Persson et al., 2007). We do have one (unplanned) experiment at sea: large-scale fishing operations have fundamentally altered marine communities over the past half century. And even better: where observations exist, we can see how marine ecosystems have responded to the removal of biomass. While these two examples provide some experimental support, most theoretical predictions stand without direct observational support.

The lack of access to controlled experiments is not unique to ecology. That challenge is shared by much of earth science, and astronomers can hardly experiment with stars. Does the lack of an active dialectic between theory and experiments make theory moot? Not quite, but it places a heavier burden upon the development of theory. As I mentioned earlier, theory is built on axioms, fundamental assumptions on which the theory rests. Theoretical physics largely rests on an agreed-upon set of axioms—Newton's laws of motion, the laws of thermodynamics, and so on—and the role of theory is making predictions on the basis of these axioms. In the subdisciplines of physics where experiments are difficult—for example, astronomy and much of earth science—the existence of these well-established laws of nature provides a solid foundation. In ecology, very few such axioms exist, and where they exist their range of validity is much more limited than the fundamental laws describing the dead nature. Ecology does not have the equivalent of Newton's laws or a Schrödinger equation to build upon. A large part of any ecological theory is therefore establishing the axiomatic foundations for the theory.

The difficulty of making experiments and direct observations of marine fish communities means that models have a special status. Model output represents our best understanding of nature. For example, fisheries management relies upon assessments of stock biomass and recruitment that are not direct observations but output of statistical models. In a similar vein, the reference points used for fisheries management, F_{msy}, F_{lim}, and so on, are not observations but are based upon model calculations. Even observations of growth rates are not directly observed but are fits to a particular growth model. In practice, however, we use such model outputs as if they were direct observations. In this manner, the models transgress from being descriptions of reality to becoming the reality itself. The lack of direct observations to check the models puts a particular burden on building trust in the models' foundational assumptions.

1.2 HOW TO READ THIS BOOK

This books presents the size- and trait-based framework for fish populations and communities as a single coherent theoretical framework (fig. 1.2). The theory is a synthesis of work over more than a decade published in more than 25 journal papers. Some of these papers are riddled with typos (for example, Andersen and Beyer, 2006), and some (if not most) are hard to penetrate (see Andersen et al., 2015, for a good example). The dense writing partly reflects the challenges in communicating complex concepts but also that my understanding was not yet fully formed while the theory was still developing. Further, the notation and some

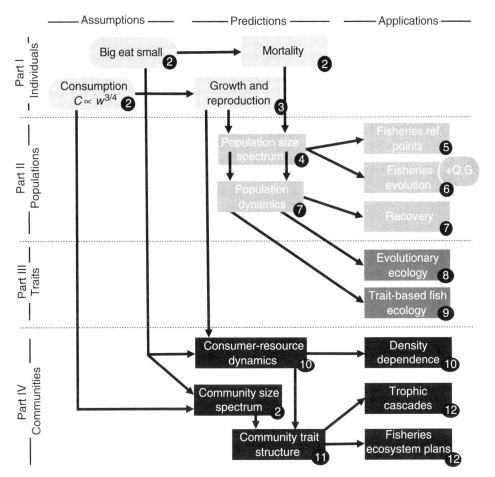

FIGURE 1.2. Sketch of the size- and trait-based theoretical framework. Boxes with rounded corners represent assumptions; chapter numbers are shown in the black circles. Fisheries-induced evolution, as addressed in chapter 6, needs further assumptions about quantitative genetics (Q.G.). Notice that the entire theory is based upon the two fundamental assumptions in the top-left corner, either directly or through concepts derived from those assumptions.

assumptions morphed throughout the process. Here, the theory is presented as a unified framework with consistent notation (summarized in table 0.1) and applied to fisheries problems, to evolutionary ecology, and to population ecology.

The size- and trait-based approach is appealing in its conceptual simplicity, but it comes at a cost of a mathematical formalism that is unfamiliar to most ecologists and fisheries scientists. I have tried to be accessible to biologists who know what an integral is but are not necessarily able to evaluate one. I do not expect prior familiarity with partial differential equations. I have focused the text on developing concepts, principles, and explaining results. Complicated mathematical derivations break the flow of reading and thinking, and consequently I have delegated them to boxes scattered throughout the text. The book can (and should) be read without going through the boxes in detail. The boxes are provided for reference and can be consulted whenever needed. All the code for the figures has been written in R. It is available at press.princeton.edu/titles/13516.html, including a Web application to simulate the impact of fishing on a stock.

The book is divided into four parts (as shown in fig. 1.2): "Individuals," "Populations," "Traits," and "Communities." Part I lays down the axiomatic foundations for the theory. The theory is rooted in assumptions at the level of individual organisms about their physiology, metabolism, clearance rate, and predator-prey interaction with smaller organisms. From that basis follows the size-structure of the entire marine ecosystem (chapter 2, "Size Spectrum Theory"). The assumptions are used to develop descriptions of how individuals grow and reproduce (chapter 3, "Individual Growth and Reproduction").

In part II, "Populations," I develop the demography of fish populations and with applications to single-stock fisheries management. By *demography*, I mean the distribution of small and large individuals within a population, which is described by the *population size spectrum* (chapter 4, "Demography"). The population size spectrum follows directly from the assumptions about growth and reproduction from chapter 3 and mortality from chapter 2. The derivation of the population size spectrum is followed up by descriptions of the ecological and evolutionary impacts of fishing (chapter 5, "Fishing"; and chapter 6, "Fisheries-Induced Evolution"). Well-established fisheries concepts such as maximum sustainable yield, yield-per-recruit, cohort biomass, and selectivity are recalculated to reveal insights hidden from classic age-based theory. The application of trait-based calculations provides broad predictions for fish stocks in general. Next, the theory is applied to population dynamics where the population changes over time, owing to environmental noise, fishing, or both (chapter 7, "Population Dynamics").

Part III, "Traits," turns away from fisheries demography and applies the theory to fundamental evolutionary problems relevant for fish (chapter 8, "Teleosts versus Elasmobranchs"). Traits represent a recurring theme, which resonates with increasing force throughout the book. The tension is released in chapter 9, "Trait-Based Approach to Fish Ecology," where I outline the conceptual mechanistic trait-based framework and link it to classic life-history theory and evolutionary ecology.

Part IV, "Communities," scales from single populations to entire communities. First, the focus is on a generalization of a classic consumer-resource model with a single population embedded in a community in chapter 10, "Consumer-Resource

Dynamics." Next, chapter 11, "Trait Structure of the Fish Community," derives the trait structure of the community. In chapter 12, "Community Effects of Fishing," I use the community model to repeat many of the classic impact calculations of a single stock on the entire community. Here, a focus is the appearance of trophic cascades. I discuss the relevance to the emerging ecosystem approach to fisheries management. Last, in chapter 13, "Opportunities and Challenges," I outline four future research questions where the theory could be applied: stochasticity, behavioral ecology, coupling to primary production, and thermal ecology and climate change.

This book does not have to be read from the start to the end. The chapters do follow a logical progression in complexity and build upon one another, but I have tried to make each chapter as self-contained as possible. This entails some repetition. I use references to previous chapters to provide links to the more fundamental chapters, like the arrows in fig. 1.2, but each chapter can be read independently. Which parts of the book you will focus on depends on your interests and background. If you are mostly interested in the fisheries applications, you might want to focus on parts II and IV, particularly the specific applications to fishery, chapters 5, 6, and 12. Perhaps you might want to consult chapter 10 for a deeper discussion of density dependence and a peek into the future of fisheries population modeling. If your interests are rather in population or community ecology, you might find the static demographic calculations in part II too altmodisch and will skip straight from part I to part IV and consult chapter 4 only for reference. However, to communicate with fisheries scientists, familiarity with the concepts in chapters 4 and 5 are essential. You might also want to read chapter 9 for inspiration about trait-based concepts in population and community ecology. In any case, I urge you to read at least the first part of chapter 2 to understand the basic assumptions of the theory, and perhaps also chapter 3. In short, follow your interest.

PART I
INDIVIDUALS

CHAPTER TWO

Size Spectrum Theory

What is the abundance of organisms in the ocean as a function of body size? If you take a representative sample of all life in the ocean and organize it according to the logarithm of body size, a remarkable pattern appears: the total biomass of all species is almost the same in each size group. The sample of marine life does not have to be very large for the pattern to appear. Every year, students from my institute venture out in a small research vessel on Øresund, the narrow straight between Denmark and Sweden, to take samples with plankton nets. When they analyze their samples in the laboratory, the pattern of biomass being roughly independent of body size consistently emerges. What is even more surprising is that the pattern extends beyond the microbial community sampled by plankton nets—it persists up to the largest fish, and even to large marine mammals.

This regular pattern is often referred to as the *Sheldon spectrum* in deference to R. W. Sheldon, who first described it in a groundbreaking series of publications. Sheldon had gotten hold of an early Coulter counter that quickly and efficiently measured the size of microscopic particles in water. Applying the Coulter counter to microbial life in samples of coastal sea water, he observed that the biomass was roughly independent of cell size among these small organisms (Sheldon and Parsons, 1967). And he saw the pattern repeated again and again when he applied the technique to samples from around the world's oceans. Mulling over this result for a few years, he came up with a bold conjecture (Sheldon et al., 1972): the pattern exists not only among microbial aquatic life, but it also extends all the way from bacteria to whales, perhaps with a small decline in the biomass for large organisms, as shown in fig. 2.1. This was a very bold conjecture, but it turned out to be largely correct. Fig. 2.2 shows Sheldon size spectra measured in four ecosystems spanning from unicellular plankton to fish (see, for instance, Sprules and Barth [2016] for more examples). In broad terms, the slight decline in biomass is evident; however, a finer undulating structure is also evident. Those undulations are partly a reflection of the difficulties of patching together measurements with different methods—plankton caught with nets of different mesh size and fish with trawl. However, they may also reflect trophic cascades. For now, I will focus on the broad pattern of biomass being a power-law function of size as indicated by

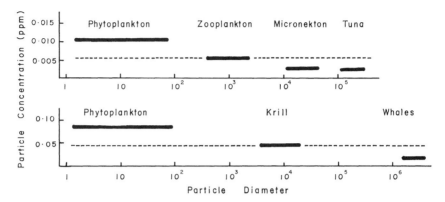

FIGURE 2.1. The standing stock of biomass in the equatorial Pacific (top) and Antarctic (bottom) as estimated by Sheldon et al. (1972). The particle diameter is measured in microns. Reprinted with permission from Sheldon et al. (1972).

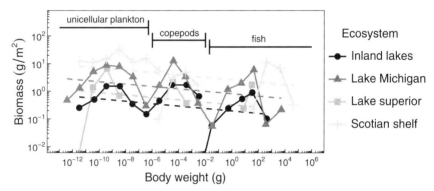

FIGURE 2.2. Observations of size spectra, from phytoplankton to fish, represented as Sheldon spectra (box 2.1). Each data point represents the biomass per area among all organisms within a factor of 10 mass range—that is, from 1–10 g, 10–100 g, and so on. The spectra were fitted to power-law functions with a common exponent ($\lambda = -0.043 \pm 0.022$). Data from Boudreau and Dickie (1992).

the fits in the fig. 2.2, but I will return to the finer structure and trophic cascades in part IV.

Sheldon's discovery did not emerge out of nowhere. The idea that the size of organisms was a key to uncover the structure of ecosystems had deep roots in evolutionary and ecological thinking. The evolutionary thinker Haldane (1928) wrote a short essay "On Being the Right Size," in which he eloquently argued how body size constrains evolution and thus determines organisms' body shape and life-history strategy. Within ecology, Elton (1926) in his classic *Animal Ecology*, stated that "[s]ize has a remarkably great influence on the organisation of animal

---

BOX 2.1

REPRESENTATIONS OF THE SIZE SPECTRUM

Three representations of the size spectrum are commonly used: the number spectrum $N_c(w)$, the biomass spectrum, and the Sheldon spectrum. The number spectrum represents the number of individuals in the size range from w_1 to w_2 as

$$\mathcal{N} = \int_{w_1}^{w_2} N_c(w)\,\mathrm{d}w. \tag{2.1}$$

Multiplying by the weight gives the biomass in the size range

$$\mathcal{B} = \int_{w_1}^{w_2} N_c(w)w\,\mathrm{d}w. \tag{2.2}$$

In a short size interval, Δw, the number of individuals is well approximated simply as $\mathcal{N} = N_c(w)\Delta w$ and likewise for the biomass.

The Sheldon spectrum can be derived from the abundance size spectrum as the biomass in a size range from w to cw, where c is the factor that determines the width of the bins. Thus, $c = 2$ is the octave bins that Sheldon used; $c = 10$ would be a base-10 grid as used in fig. 2.2:

$$\mathcal{B}_{\text{Sheldon}} = \int_w^{cw} N_c(\omega)\omega\,\mathrm{d}\omega, \tag{2.3}$$

where ω is a dummy variable for the integration. Inserting the power-law spectrum from eq. 2.5, $N_c = \kappa_c w^{-\lambda}$ gives

$$\mathcal{B}_{\text{Sheldon}} = \kappa_c \frac{c^{2-\lambda} - 1}{2 - \lambda} w^{2-\lambda} \propto N_c(w)w^2. \tag{2.4}$$

All terms except $w^{2-\lambda}$ are independent of size; hence, the Sheldon spectrum is proportional to the number density spectrum multiplied by body weight squared.

---

communities." He then went on to describe biomass "pyramids," with the biomass of small organisms at the base of a pyramid and each successive layer representing biomasses of higher trophic levels. Sheldon took Elton's biomass pyramids to the aquatic ecosystems, rebranded them "size spectra," and used them to reveal some of the hidden order in the confusingly complex marine ecosystems.

The existence of the Sheldon spectrum raises two immediate questions: Why does this pattern occur so consistently? And what does it tell us about the organization of marine ecosystems? Clearly, a very robust explanation is needed that does not rely on the specifics of the ecosystem or the organisms inhabiting it—the explanation should even span across realms of life in the oceans, from bacteria over phytoplankton, zooplankton, fish, sharks, and to whales. This chapter answers these two questions.

Sheldon's insight relied on his willingness to ignore which species the individual organisms belong to. Circumstances dictated that choice: Sheldon's Coulter counter reported only the number of particles with a given size and provided no information about the species. For an ecologist, that must have been frustrating, because the species concept plays the lead role in population ecology. In terrestrial systems, where organisms are larger and it is easy to observe their species, body-size relations are often represented as species abundance relationships, showing the abundance of populations as a function of the typical body size of individuals in the species. Even though a description of species biomass as a function of body size appears similar to the biomass spectrum, it is something quite different. If, for example, the biomass or abundance of species is plotted as a function of their body size (or some characteristic measure of body size), the beautiful regularity of the Sheldon spectrum will stay hidden (Jennings, 2007; Reuman et al., 2008). Sure, there will be some hint of a pattern; species with small body size are typically more abundant than species with large body size, but rare species exist among species with all body sizes, which muddles the picture. Only when the individuals are lumped together across species does the regularity of the Sheldon spectrum emerge. If Sheldon had access to a counter that could sort individuals according to their species, he would probably have plotted species abundances, and he might not have discovered the Sheldon spectrum.

I will show how the Sheldon spectrum emerges from predator-prey interactions and the limitations that physics and physiology place on individual organisms. How predator-prey interactions and physiological limitations scale with body size are the central assumptions in size spectrum theory, and it is therefore necessary to devote some space to setting these concepts straight. First, I'll define *body size* and *size spectrum*. Next, I'll show how central aspects of individual physiology scale with size: metabolism, clearance rate, and prey size preference. On that basis, it is possible to derive a power-law representation of the size spectrum by considering a balance between the needs of an organism (its metabolism) and the encountered prey, which is determined by the spectrum, the clearance rate, and the size preference. Last, I use the solution of the size spectrum to derive the expected size scaling of predation mortality.

2.1   WHAT IS BODY SIZE?

The term *body size* is a fairly unspecific reference to any measure that characterizes an organism's size, be it length, surface, volume, or a measure of mass or weight. There are two categories of body size measures: measures of physical

TABLE 2.1. Commonly Used Conversions Between Representations of Body Size of Fish.

	To Wet Weight	From Wet Weight
Body length[a], l	$0.01l^3$	$(100w)^{1/3}$
Dry weight[b], w_{dw}	$4.62\,w_{dw}$	$0.22\,w$
Carbon weight, w_C	$8\,w_C$	$0.125\,w$

Note: Wet weight w is measured in grams and lengths l in centimeters.
[a]Froese (2006) determines the relation between the coefficient c and exponent b in the weight-length relationships as $\log_{10}(c) = 4.544 - 2.174b$ with the median value of $b = 3.025$.
[b]Boudreau and Dickie (1992).

size (diameter, length, volume) that characterize the physical extent of the body, and measures of body mass or weight (wet weight, carbon weight, or dry weight). Physical and mass measures are related to one another through body density and shape. Whether one or the other measure of body size is used is mostly a matter of convenience, and what is most convenient depends on the specific question. Physical size is most convenient for physical and physiological measures—for example, the body surface scales with linear size, which determines fluid mechanical drag or the uptake surfaces for oxygen (gills) or food (the gut). Physical size also largely determines preferred predator-prey size ratio—an organism prefers prey of a given size relative to its gape. Body mass is a convenient measure to represent the amount of energy contained in a organism. An important aspect of mass is that it is conserved: the carbon consumed by an organism will be either accumulated, excreted, shed as reproduction, or burned and respired. Since the core of the size-based models is an energy and mass budget, I will use body mass as the fundamental measure of body size.

Body mass can be measured as dry weight, wet weight, carbon mass, or nutrient mass—for example, nitrogen or phosphorus content. The nutrient mass is often used for primary producers that are mostly limited by nutrients. However, higher organisms use energy in the form of carbon and actually excrete excess nutrients. I will therefore use body mass in terms of wet weight with the symbol w. When conversions between length l and mass are needed, I use the standard conversion: $w = cl^3$, with $c = 0.01$ g/cm^3, noting a substantial variation in both constant and exponent between different body shapes of fish (Froese, 2006). Common conversions between measures of body size are given in table 2.1.

2.2 WHAT IS A SIZE SPECTRUM?

The size spectrum represents the abundance or biomass of organisms as a function of their size. Abundance and biomass can be measured in two types of unit: either as the total abundance or the total biomass in a system, or as a concentration measure. In the first case, the units are just numbers or biomass; in the second case, units are numbers or biomass per area or per volume. Whether one uses total abundance and biomasses or a concentration measure depends on the application. For the theoretical developments in the following, it makes most sense to work with concentrations (abundance or biomass per volume), and I will do so implicitly. Therefore, references to abundance or biomass are concentrations, even though I do not write them explicitly.

Sheldon represented the size spectrum as a histogram of biomass in logarithmically spaced body-size bins—for example, from 2 to 4 g, from 4 to 8 g, and so on. That representation clearly showed how biomass is roughly independent of size, but is an inconvenient representation of the absolute level of biomass in an ecosystem. The problem is that the level of biomass in each bin depends on the bin width. If bin widths are increased—for example, using 1 to 10 g, 10 to 100 g and so on—the biomass in each bin will increase correspondingly. Representations showing different levels of biomass in different ecosystems, like the four examples in fig. 2.2, are therefore difficult to compare between publications because they are made with different bin widths. A representation that is independent of bin width is needed. This is acheived by dividing the biomass in each bin with the bin width. This procedure turns Sheldon's histogram into the biomass spectrum $B_c(w)$, with dimensions of biomass per body mass. The subscript c on B_c is used to emphasize that the spectrum represents the entire community in contrast to the population size spectra to be introduced in chapter 4 that represent populations. We can further create the number spectrum by dividing the biomass spectrum with body size: $N_c(w) = B_c(w)/w$. The dimensions of the number spectrum is numbers per body mass. Taken together, we now have three representations of the size spectrum: the Sheldon spectrum, which is Sheldon's histogram of biomasses, the biomass spectrum $B_c(w)$, and the number spectrum $N_c(w)$ (fig. 2.3). Whereas the Sheldon spectrum will vary in level depending on the bin width, the two latter representations are independent of bin width—they are also commonly referred to as "normalized" size spectra (Sprules and Barth, 2016).

The biomass and number spectra do not directly show the biomass or abundance of individuals with a given size. To recover the biomass or abundance requires a specification of a range of body sizes (the *bin width*) (see box 2.1). The biomass in a small body size range Δw is $B_c(w)\Delta w$, and likewise the abundance in that range is $N_c(w)\Delta w$. This may seem odd, but turns out to be quite convenient, as integrals over a size spectrum directly provide the abundance or biomass in a size range. It also provides a simple procedure to derive the size spectrum from binned data: the size spectrum emerges by dividing the abundance or biomass in each bin by the bin width (see White et al., 2007; Edwards et al., 2017, for more elaborate statistical methods).

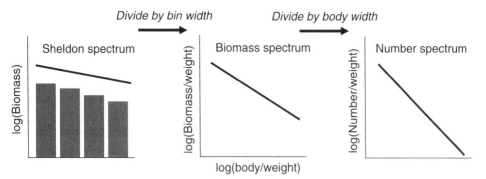

FIGURE 2.3. Three representations of size spectra: the Sheldon spectrum, the biomass spectrum, and the number spectrum. The arrows indicate transformations between the representations. See box 2.1 for details.

The size spectra in fig. 2.2 can be approximated with power-law functions. The number or biomass spectra can be written as

$$N_c(w) = \kappa_c w^{-\lambda} \quad \text{or} \quad B_c(w) = \kappa_c w^{1-\lambda}, \tag{2.5}$$

where κ_c is the coefficient (often referred to as the *intercept*) and λ the spectrum exponent (*slope*). The Sheldon spectrum is proportional to the biomass spectrum multiplied by the body size: $B_{\text{Sheldon}} \propto B_c(w)w \propto w^{2-\lambda}$ (see box 2.1). If the Sheldon spectrum is flat—that is, biomass is independent of body size—the exponent is zero, and thus $\lambda = 2$. We therefore expect that the number spectrum has an exponent around -2 and the biomass spectrum an exponent around -1. The coefficient κ_c representing the total biomass in a system will vary between systems, as shown in fig. 2.2.

2.3   SCALING OF PHYSIOLOGY WITH BODY SIZE

The remarkable regularity of the Sheldon spectrum is a result of interactions between the fundamental units in the ecosystem: the individual organisms. All pelagic multicellular organisms fuel their metabolism by burning the carbon contained in other, smaller, organisms. The organisms can be likened to the molecules in a gas, flying around and interacting through inelastic collisions. The interactions between the organisms in an ecosystem are the predator-prey encounters. Here, the analogy with the molecules in the gas break down: in a interaction between two molecules, both molecules survive the encounter, while in a predator-prey encounter only the predator survives.

The predator-prey encounter process is described by three quantities: the clearance rate, the maximum consumption rate, and the prey size preference function. The three quantities measure the predator's ability to catch prey, its ability to process the prey, and which size of prey it prefers. How these processes scale with body size is constrained by fundamental physics. The clearance rate is limited by the swimming speed and the sensing range of the predator; the maximum consumption rate is limited by the surface that absorbs the prey; and the prey size preference is constrained by the predator's ability to catch and handle different-size prey. In this section, I combine theoretic arguments and fits to empirical data to establish the size scalings of these three relations.

Clearance Rate

A predator's ability to hunt for prey is described by the clearance rate. It has dimensions of volume per time—for example, liters per day. It is best understood by reference to a filter-feeding predator that filters a volume of water per day. The filtered volume is proportional to the filter area multiplied by the cruising velocity. The larger the predator, the larger the filter, and we can assume that the diameter increases proportional to the length of the predator. The filter area is then proportional to the length of the predator squared, l^2. The cruising velocity of a microbial organism, such as a flagellate or a copepod, is roughly proportional to the length to the power 0.75 (Andersen et al., 2015). The volume cleared is therefore proportional to $l^{2.75}$, or, using the length-mass relationship $w \propto l^3$, to $w^{0.92}$. The clearance rate can therefore be described as a power law:

$$V(w) = \gamma w^q, \tag{2.6}$$

with exponent q and coefficient γ. Not many predators, though, are true filter feeders. Instead, they encounter their prey while actively searching for it or by waiting for the prey to come to them. Nevertheless, their encounter with prey can still be characterized as an effective clearance rate. In this case, the clearance rate is the encounter rate of organisms (numbers per time) divided by the concentration of prey (numbers per volume). The clearance rate can be inferred as the slope at origin of the functional response that measures the consumption rate as a function of prey concentration. Observed clearance rates are fairly well described as a power-law function with an exponent close to 1 (fig. 2.4). If only fish are considered, the exponent is smaller—the fits gives $q \approx 0.76$—though it should be noted that there are no values of fish larger than 100 g, so the value is fairly uncertain. The indicated smaller exponent for fish than for microbial organisms corresponds to a smaller exponent of the scaling of cruising velocity of fish, roughly $\propto l^{0.45}$ versus $\propto l^{0.75}$ for microbial organisms (Andersen et al., 2015). As this book is mainly concerned

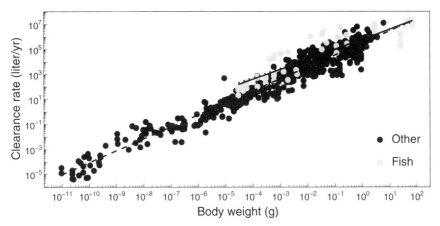

FIGURE 2.4. Clearance rate as a function of body weight for organisms ranging from nano-flagellates to fish (Kiørboe and Hirst, 2014). Fits are made to all groups (dashed line), to fish only (dotted line) and to fish using exponent $q = 0.8$ (solid line). Coefficients for the fitted values are summarized later in this chapter, in table 2.2.

with fish, I define a "canonical" value of the exponent that is smaller than the fit to all marine life but slightly larger than the fit to data for fish—namely, $q = 0.8$.

Respiration and Consumption Rate

A predator's food consumption should at least satisfy its basal metabolic requirements for survival. At the same time, the consumption cannot exceed the maximum digestive capacity. In this way, the standard metabolism and the maximum consumption rate describe the organism's metabolic requirements and capacity. Both quantities clearly follow power laws with almost similar exponents, with a value less than 1 (fig. 2.5). This scaling exponent reflects the celebrated Kleiber's law, which states that standard metabolism scales with body mass with an exponent roughly equal to $3/4$ (Kleiber, 1932; West et al., 1997). The scaling exponent emerges because both processes are limited by transport of oxygen, nutrients, and carbon through surfaces in the body—for example, the gills and the digestive tract. The surface of an organism is proportional to the length squared, or the weight to exponent $2/3$, and we would then expect the exponent of metabolism and maximum consumption to have exponent $2/3$. However, as already pointed out by Haldane (1928), the surface of lungs, gills, and digestive tracts are not regular, but are instead folded.[1] The surfaces are fractal, and their surface area therefore scales

---

[1] Not only did Haldane provide the gifted insight that the 2/3 law is too simple for describing how surfaces limit uptake, his writing is also a joy in itself: "When a limit is reached to their absorptive powers their surface has to be increased by some special device. For example, a part of the skin may

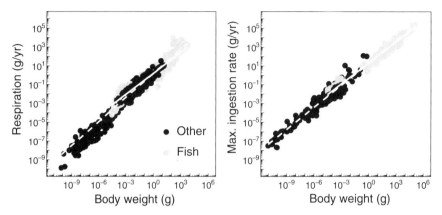

FIGURE 2.5. Respiration and maximum ingestion rates as functions of body weight for organisms ranging from nanoflagellates to fish. Data of nonfish and larval fish from Kiørboe and Hirst (2014). Data for fish are based on fits to observed growth curves of adult fish with a procedure to be described in chapter 3 (eqs. 3.10 and 3.31). The weight selected for each observation is the size at maturation. Fits are made to all groups (dashed line), to fish only (dotted line) and to fish using exponent $n = 3/4$ (solid line).

with an exponent larger than $2/3$. An argument has been put forward that $3/4$ is the optimal dimension of a fractal delivery network (West et al., 1997). Though the argument about optimal fractal dimension is contested, there is no doubt that the scaling exponent is higher than $2/3$, and I will here use the canonical value of $3/4$. The maximum consumption is then

$$C_{\max} = hw^n, \tag{2.7}$$

where $n = 3/4$ is the *metabolic* exponent.

Prey Preference

Predators eat prey smaller than themselves. Stomach samplings show that the size of prey in the stomach of predators is roughly 700 times smaller in mass than the predator (fig. 2.6). However, we need to know what size of prey the predator would prefer to eat if it could choose freely, and that is not necessarily the same as the sizes of prey the predator has eaten. What size of prey the predator prefers cannot be observed directly. If, for example, a predator would like to eat prey of

---

be drawn out into tufts to make gills or pushed in to make lungs, thus increasing the oxygen-absorbing surface in proportion to the animal's bulk." While Haldane knew that the scaling $2/3$ was incorrect, he seems to conclude that the convoluted nature of the surfaces makes them scale linearly with body weight ("in proportion to the animal's bulk"), which is also incorrect.

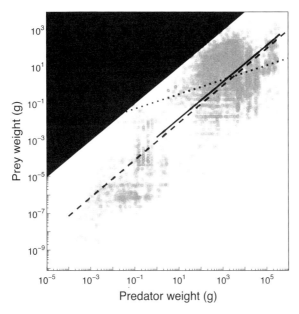

FIGURE 2.6. Sizes of prey in the stomach of ectotherm vertebrates as a function of weight. Fits to all observations (dashed line), to larger organisms (predominantly fish; dotted line), and to large organisms with fixed exponent 1 (solid line). The dark gray area indicates that prey are larger than the predator. Data are from Barnes et al. (2008).

10 g but there is mainly prey of 1 g available, then the stomach contents will be dominated by the 1 g prey. Inferring the preference requires knowledge of which sizes of prey the predator had to choose from. The Sheldon spectrum shows us that the biomass of prey is roughly independent of size, which again means that smaller prey are more abundant than larger prey—the 1 g prey would be 10 times more abundant than the 10 g prey. Even though the literature abounds with measurements of stomach contents, as shown in fig. 2.6, there are very few attempts to measure the preference for prey size. A pioneering effort was by Ursin (1973), who examined the stomach of cod and dab and corrected for the abundance of prey. He described the preference of prey of size w_p by a log-normal prey size preference function (fig. 2.7)

$$\phi(w, w_p) = \exp\left[\frac{-(\ln(w/(\beta w_p)))^2}{2\sigma^2}\right], \tag{2.8}$$

where β is the preferred predator:prey mass ratio and σ the width of the size selection function. Ursin showed that the preferred predator:prey mass ratio was generally smaller than the predator:prey size ratio of prey in the stomach, and he

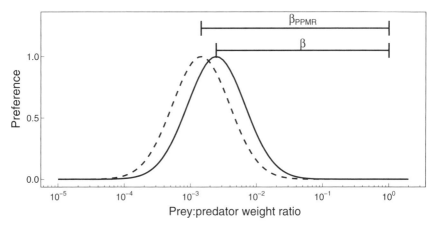

FIGURE 2.7. The predator-prey size preference function (eq. 2.8) drawn with the prey:predator ratio w/w_p on the x-axis. The solid line shows the preference for prey with β being the preferred predator:prey size ratio. The dashed line shows the distribution of prey in the stomach, with β_{PPMR} being the average predator:prey size ratio of consumed prey, which is the quantity plotted in fig. 2.6.

found that the preferred predator:prey size ratio β was around 100 for cod and 1000 for dab.

Knowing the preference of predators toward prey is a key component of multispecies ecosystem models. To this end, the countries around the North Sea established two massive samling programs, referred to as the "years of the stomach," where tens of thousands stomachs were cut open and the more or less digested contents carefully counted and measured. By comparing the stomach contents with prey abundances established from concurrent trawl samples, the preference for prey species and sizes were estimated (Lewy and Vinther, 2004). Analysis of the massive database showed a variation of the preferred predator:prey size ratio between around 200 and 2000, with a geometric mean of 469. An alternative estimate can be made from observed predator:prey ratio in the stomachs from fig. 2.6 corrected for the expected abundance of prey under the assumption that the prey distribution follows a Sheldon spectrum. Hartvig et al. (2011) calculated the correction factor between the predator:prey mass ratio in the stomach and the preference to be 1.7 (see also box 2.2). Correcting the observed predator:prey mass ratio in the stomach of fish of around 708 gives a value of $\beta \approx 400$, right between the two values observed by Ursin for cod and dab, and in accordance with the estimates from the North Sea stomach sampling campaigns.

The three individual-level properties—the clearance rate (eq. 2.6), the maximum consumption rate (eq. 2.7), and the prey-size preference function (eq. 2.8)—together describe the main aspects of the interactions between prey and predators

TABLE 2.2. Fits to Physiological Rates from section 2.3 (Top) and Derived Quantities from sections 2.4 to 2.7 (Bottom).

Process	All*	Fish Only	Canonical Exponent
Physiological rates			
Clearance rate $V(w)$ (liter/yr)	$3.3 \cdot 10^5 \, w^{0.96}$	$6.0 \cdot 10^5 \, w^{0.76}$	$1.2 \cdot 10^5 \, w^{0.8}$
Maximum ingestion (g/yr)‡	$17.2 \, w^{0.77}$	$17.6 \, w^{0.76}$	$16.0 \, w^{3/4}$
Respiration (g/yr)‡	$1.65 \, w^{0.85}$	$1.7 \, w^{0.78}$	$1.18 \, w^{3/4}$
Pred:prey in stomach β_{PPMR}	$1224 \, w^{1.02}$	$\dagger$	$708 \, w^1$
Spectrum $\kappa_c w^{-\lambda}$	$8.82 \cdot 10^{-6} \, w^{-2.11}$		$2.35 \cdot 10^{-5} \, w^{-2.05}$
Derived quantities			
Preferred predator:prey $\beta^\S$	665		408
Abundance factor Φ_a	3.54		3.39
Predation factor Φ_p	0.07		0.17
Trophic efficiency ε_T	0.07		0.16

Note: All units in grams, liters, and years.

*Spectrum in the "All" case calculated under the assumption of a predator-prey size preference of 1224 independent of body size.

$\dagger$ The fit is poor (see fig. 2.6) and the values are therefore not considered useful. This also means that values of the parameters that rely on the observed predator:prey mass ratio in that column cannot be estimated.

$\ddagger$ Maximum ingestion and respiration is fitted with the same exponent.

$\S$ Calculated as $\beta = \beta_{PPMR}/1.7$.

in marine ecosystems. I have now established the values of the exponents and the coefficients for these relations. The values are summarized in the first part of table 2.2. Those values will be used to obtain quantitative estimates from the theory to be developed in the following chapters. Nevertheless, the exact values of the exponents and coefficients are not crucial for the qualitative results of the theory. If, for example, the exponent of the clearance q is 1, or if the metabolic exponent is $2/3$, the framework still stands solid. What matters somewhat, though, is that these relations are well described by power-law relations. That this is indeed the case was made abundantly clear by the data analyses. In terms of values of the exponents, I define "canonical" values of $q = 0.8$ for the clearance rate, $n = 3/4$ for standard metabolism and maximum consumption, and 1 for prey size preference (such that preferred prey size is directly proportional to predator size) in the numerical examples that follow. However, the formulas given are general and other values of the exponents can be used.

The statistically inclined reader may have looked for confidence intervals and measures about goodness of fits. I have omitted this information to emphasise that the exact numerical values of exponents and coefficients are not central for the qualitative results. Had I provided uncertainties, p-values and R^2 measures,

BOX 2.2

INTEGRALS OVER THE SIZE SPECTRUM

The derivation of the size spectrum in section 2.4, the mortality in section 2.5, and the size of prey in the stomach all rely on integrals over the size spectrum.

The biomass of encountered prey by a predator of size w is found by integrating over all prey sizes w_p (eq. 2.19)

$$B_{\text{prey}}(w) = \int_0^\infty N_c(w_p) w_p \phi(w/w_p) \, dw_p. \tag{2.9}$$

Inserting the power-law form of community size spectrum $N_c(w_p) = \kappa_c w_p^{-\lambda}$, this integral can be solved by laborious calculations

$$B_{\text{prey}}(w) = \Phi_a \kappa_c w^{2-\lambda}, \tag{2.10}$$

where

$$\Phi_a = \sqrt{2\pi} \sigma \beta^{\lambda-2} \exp[(\lambda-2)^2 \sigma^2/2]. \tag{2.11}$$

The constant Φ_a involves the parameters in the predator-prey preference function (eq. 2.8). It is proportional to width of the preference σ—clearly, more prey is available if the predator is able to eat a wide size range of prey. Φ_a also involves the predator:prey mass ratio; however, this is raised to a small exponent ($2 - \lambda$ is close to zero), and thus the dependency is weak. This is not surprising; since the Sheldon spectrum is almost flat, the biomass of available prey is roughly independent of the size of prey. The value of $\Phi_a \approx 3$ (table 2.2).

The mean size of prey in the stomach is

$$
\begin{aligned}
w_{\text{prey}} &= \frac{\int_0^\infty N_c(w_p) w_p \phi(w_p/w) \, dw_p}{\int_0^\infty N_c(w_p) \phi(w_p/w) \, dw_p} \\
&= \frac{B_{\text{prey}}}{N_{\text{prey}}} = \sigma e^{(3-\lambda)/(2\sigma^2)} w/\beta.
\end{aligned}
\tag{2.12}
$$

The ratio between the weight of the predator and consumed prey is

$$\beta_{\text{PPMR}} = w/w_{\text{prey}} = \beta e^{(\lambda-3)/(2\sigma^2)}/\sigma. \tag{2.13}$$

For default parameters values, $\beta_{\text{PPMR}} \approx 707$. The consumed prey is therefore a factor $\beta_{\text{PPMR}}/\beta \approx 1.7$ smaller then the preferred weight.

The mortality inflicted on prey is the fraction of the volume cleared by predators per time, weighted by the size preference of the predators. The volume cleared by a predator of size ω is $V(\omega)$. To get the fraction cleared by all predators, the

(continued)

(Box 2.2 *continued*)

clearance rate should be multiplied by the density of predators $N_c(\omega)$ and integrated. The mortality on prey of size w is then

$$\mu_p(w) = \int_0^\infty V(\omega)N_c(\omega)\phi(\omega/w)\,d\omega. \tag{2.14}$$

Inserting the solution of the size spectrum from eq. 2.21 gives

$$\mu_p(w) = f_0 h \Phi_p w^{n-1}, \quad \text{with} \quad \Phi_p = \beta^{2n-q-1} e^{(2n-q-1)(q-1)\sigma^2/2}. \tag{2.15}$$

The constant Φ_p again involves the predator-prey parameters, and has a value in the range 0.1 to 0.2 (table 2.2).

they would generally have been very favorable (except in the case of the predator:prey mass ratios for fish, where I have refrained from providing fits). Good values of statistical fits may lull us into false complacency and trust in the data, and make us forget that obtaining accurate values are about more than statistics. All of the data sets are fraught with potentially systematic errors related to the actual measurement: they are typically performed in artificial laboratory settings where behavior may well be different from that in situ; stomachs might be regurgitated before they are analyzed, and differential digestion rates of different-size prey is inaccurately represented (if at all); respiration measurements ignores potentially significant contributions from activity, and so on. Last, there is the issue of selection of data points. The meta-analyses that I mined for data are biased toward species of commercial value in temperate waters, such as cod, herring, and so on,[2] or very diverse groups with high conservation value, such as rockfish, and are therefore not likely to be a representative selection of fish life histories. While keeping these issues in mind, it is clear that the power-law function is a good model to describe the data, and that the estimates of the parameters represent our current best knowledge of the predator-prey interactions.

While some of the power-law fits are impressive, it should be remembered that there are other explanatory variables in play besides body size. When power-law fits are made over a large size range—here, over 16 orders of magnitude—the power-law nature of the data clearly emerges. Still, there is a substantial variation around the mean. For example, the measurements of maximum ingestion in fig. 2.5 rates varies almost two orders of magnitude around the mean. Consequently, body

---

[2] I gravitate to use the culturally and economically important fish species from Denmark as examples, such as herring and cod, because I know them well. I could just as well have used anchovy and hake, or other examples of planktivorous and piscivorous species.

size is a fairly poor predictor of maximum consumption of organisms of similar size. Comparing, for example, a herring of 10 g and a cod of 100 g, the prediction from the size-based relationship in eq. 2.7 may be quite off the mark because the differences between the two species that are not described by body size may lead to larger differences in maximum consumption rates than the difference in size. If, however, two herrings of 10 and 100 g are compared, the purely size-based relation will be a good predictor of the difference between these two individuals, because species differences are not relevant and only the difference due to body size matters. Only when individuals with large size differences are compared are the pure size-based predictions accurate.

In the following, I will use the physiological relations that I have established to gain insight into several aspects of marine communities: the community size spectrum, the predation mortality, the length of marine food chains, and the trophic efficiency. The derivations of the size spectrum and the predation mortality are used extensively to derive other results, while the length of marine food chains and the trophic efficiency are provided for completeness.

2.4   WHAT IS THE SIZE SPECTRUM EXPONENT?

Having established the three fundamental scaling laws describing the predator-prey encounter process, we can use them to understand why the Sheldon spectrum is so commonly observed in marine ecosystems. The key is the relation between consumption C and available prey

$$C = VB_{\text{prey}}, \tag{2.16}$$

where V is the clearance rate and B_{prey} is the biomass concentration of available prey. This relation can be understood from the dimensions of the quantities: the clearance rate has dimensions of volume per time and the biomass concentration is biomass per volume. Their product has dimensions of biomass per time, which are the same dimensions as the consumption. The relation in eq. 2.16 can be used to determine the size spectrum (eq. 2.5) (Andersen and Beyer, 2006): knowing C and V we can solve eq. 2.16 for the biomass of prey

$$B_{\text{prey}} = C/V. \tag{2.17}$$

If the size-scaling of consumption and clearance rate are known, the dependency of prey biomass with size can be found—and that is exactly what the Sheldon spectrum represents.

We know the clearance rate from eq. 2.6, but we do not yet know the consumption rate. The consumption rate can be deduced from the maximum ingestion rate.

We can safely assume that the consumption is smaller than the maximum consumption rate. Consumption can therefore be written as a factor multiplied by the maximum consumption (eq. 2.4)

$$C = f_0 C_{\max} = f_0 h w^n, \tag{2.18}$$

where the "average feeding level" f_0 is between 0 and 1. f_0 should be large enough to satisfy standard metabolism, but also less than 1 to represent that fish with full stomachs are rarely caught (Armstrong and Schindler, 2011). $f_0 = 0.6$ is an appropriate choice.

All that remains is to relate the biomass of prey B_{prey} to the size spectrum $N_c(w)$. The biomass of prey for a predator of size w is the sum of the biomass of all prey w_p weighted by the prey size preference function (eq. 2.8)

$$B_{\text{prey}}(w) = \int_0^\infty N_c(w_p) w_p \phi(w/w_p) \, dw_p = \Phi_a \kappa_c w^{2-\lambda}, \tag{2.19}$$

where the power-law form of the size spectrum $N_c(w_p) = \kappa_c w_p^{-\lambda}$ has been used and the value of the dimensionless constant Φ_a is given in box 2.2. Inserting eq. 2.19, $C = f_0 C_{\max}$ and $V = \gamma w^q$, into eq. 2.17 and isolating the size spectrum gives

$$N_c(w) = \kappa_c w^{-\lambda} = \frac{f_0}{\Phi_a} \frac{h}{\gamma} w^{-2-q+n}, \tag{2.20}$$

or equivalently for the size spectrum coefficient and exponent

$$\kappa_c = \frac{f_0}{\Phi_a} \frac{h}{\gamma} \quad \text{and} \quad \lambda = 2 + q - n \approx 2.05. \tag{2.21}$$

This is a powerful result. It provides a direct relation between individual level processes—clearance rate (q and γ), consumption (n and h), and predator-prey preference (β and σ within Φ_a)—with community measures related to the size spectrum: the abundance coefficient κ_c and the exponent λ. Inserting the values of the parameters from table 2.2 into eq. 2.21 gives the values of size spectrum abundance κ_c and exponent λ. Fig. 2.8 shows that the predictions fit measurements of the size spectrum remarkably well—remember that the lines are not statistical fits, but predictions based upon the basic processes from section 2.3.

The relations in eq. 2.21 provide more than predictions with good correspondence to observations. First, the exponent is 2 with a small correction given by the difference between the exponent of maximum consumption n and clearance rate q. An exponent of 2 corresponds to a flat Sheldon spectrum (box 2.2). The exponent of the clearance rate q is larger than the exponent of the maximum consumption n, and the exponent of the spectrum is therefore somewhat larger than 2

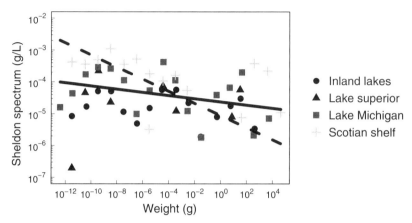

FIGURE 2.8. The Sheldon spectrum as calculated from eq. 2.21 using the fitted parameters of physiological rates for fish with the "canonical" exponents (solid line) and from the fitted values to all marine organisms (dashed line) (table 2.2). Data points are the same as fig. 2.3 converted from the "per area" concentration of abundance to a volumetric concentration by assuming a depth of the productive layer of 30 m.

and the Sheldon spectrum declines slightly with size—as is well reflected in measured spectra fig. 2.8 and by Sheldon's original conjecture fig. 2.1. Second, the spectrum abundance κ_c is the product of two ratios: the ratio between the average feeding level f_0 and the predator-prey interaction coefficient Φ_a, and the ratio between the coefficients of maximum consumption rate and clearance rate h/γ. The ratio f_0/Φ_a is on the order of 1 (actually a little smaller; see table 2.2) and only weakly dependent on the parameters β and σ, which are related to the size preference function. That ratio is therefore not very important. The second ratio h/γ is more interesting. It states that an ecosystem with a high biomass (large κ_c) will be populated by species with high consumption rates (large h) and therefore fast growth rates. Conversely, depleted ecosystems, such as deep-sea systems (low κ_c), will be populated by slow-growing species.

The relationship between the spectrum and fundamental physiological rates established in eq. 2.21 raises a somewhat philosophical question: is the spectrum determined by the physiological parameters or are the physiological parameters determined by the spectrum? Eq. 2.21 shows us how the parameters are related, but it does not offer an answer to this question. I lean toward the perspective that the abundance of the spectrum, the κ_c parameter, is determined by the amount of energy entering the system, be it from primary production as in a pelagic system or from detritus in a deep-sea system. The amount of energy in turn determines which types of species, characterized by the ratio between maximum ingestion and clearance rate h/γ, can exist in the system. The value of the exponents of the fundamental physiological relations, on the other hand, are determined by limitations set by physics and physiology and therefore independent of the environment. The spectrum exponent is therefore the same between ecosystems and does not depend on specifics, such as the productivity, the temperature, or the predator:prey size ratio.

We now have established an explanation for the Sheldon spectrum. The pattern emerges as a consequence of the predator-prey relationships between aquatic organisms governed by the rule: bigger fish eat smaller fish. But what does the Sheldon spectrum mean ecologically? We can use it to make a general rule of thumb for the number of prey needed to support a predator. In a flat Sheldon spectrum, where the biomass of prey is roughly independent of size, the biomass of predators is the same as the biomass of prey. The number of prey per predator will then depend on how much smaller the prey are than the predator and roughly given by the predator-prey mass rate. The number of prey per predator is therefore $\beta \approx 500$.[3]

The mathematical derivation of the Sheldon spectrum is *mechanistic*: it is based on robust assumptions about the system on a lower order of organization—the individuals. A mechanistic explanation is satisfying from a scientific and aesthetic perspective because it formalizes an understanding about how nature works. Mechanistic explanations, such as the understanding of the scaling of the size spectrum, have further utility than enlightenment: they can be used to make quantitative predictions and explain other observed patterns. In the next three sections, I will used the insights gained from the mechanistic explanation of the size spectrum to predict how mortality scales with body size, the length of marine food chains, and the trophic efficiency of marine ecosystems. Of these three predictions, the mortality is a key ingredient in several of the following chapters.

2.5   WHAT IS THE PREDATION MORTALITY?

Very few marine organisms die of old age—they get eaten by larger organisms before senescence takes its toll. The primary cause of mortality is therefore predation. Fig. 2.9 shows how mortality declines with body size, with an exponent roughly around -0.25 (McGurk, 1986; Hirst and Kiørboe, 2002; Gislason et al., 2010). A "metabolic" line of reasoning relates the mortality exponent to the metabolism exponent by a dimensional argument: if metabolism, with dimensions

---

[3] This rule of thumb was used by Sheldon and Kerr (1972) to estimate the number of Loch Ness monsters. They found that Loch Ness could support a population of 10 monsters. They were also careful to mention that "their most characteristic features are that they are rarely seen and never caught."

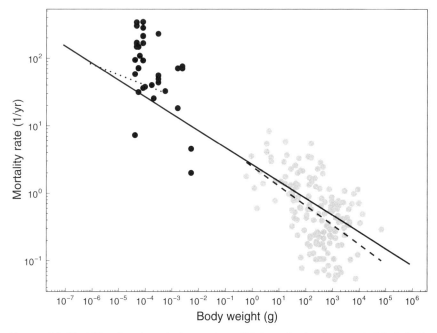

FIGURE 2.9. Mortality of marine pelagic organisms; fish (gray dots) and copepods (black dots). The black line is eq. 2.22 using the canonical parameter values from table 2.2, and the dotted and dashed lines are fits to copepods and fish, respectively. Data for post-larval fish are from Gislason et al. (2010) and for copepods from Hirst and Kiørboe (2002). All rates are corrected to 15°C with a $Q_{10} = 1.83$.

mass/time, scales as 3/4, then mass-specific rates, such as the mortality, should scale with exponent $3/4 - 1 = -1/4$ (Brown et al., 2004). This argument supplies the exponent, but it does not provide a prediction of the level of mortality nor does it describe the mechanism that regulates the scaling towards the $-1/4$ exponent. We can exploit our understanding of how the size spectrum is shaped by predator-prey interactions to derive the predation mortality (Andersen et al., 2009a). The predation mortality is derived in box 2.2

$$\mu_p(w) = \Phi_p f_0 h w^{n-1}, \tag{2.22}$$

with the constant Φ_p being on the order of 0.1 and determined by the parameters in the prey preference function (eq. 2.15). The exponent of the weight-scaling is $n - 1 = -0.25$, in accordance with the metabolic argument.

The predation mortality in eq. 2.22 adds two aspects to the simple metabolic argument. First, it makes an explicit prediction of the coefficient, which turns out to be a fairly good prediction of the observed mortality in fig. 2.9. Second, the presence of a coefficient related to consumption, $f_0 h$, explicitly links consumption and growth to mortality: higher consumption by predators (and thus faster growth) imposes as higher mortality on their prey. Though the relation between growth and mortality in eq. 2.22 is established between predators and prey, it is generally thought to be valid also for the predator itself. A predator with fast growth (high value of h) needs more food and therefore tends to expose itself to higher predation mortality and vice versa. The relation between predation mortality and growth is embodied in the M/K life-history invariant first introduced by Beverton (1992) for fish and later generalized by Charnov et al. (2001). The M refers to adult mortality, and K is a measure of the growth rate. I will return to the ratio M/K in more detail in chapter 4; for now, it suffices to notice that eq. 2.22 provides a mechanistic explanation of the M/K life history invariant.

2.6   HOW LONG ARE MARINE FOOD CHAINS?

Marine food chains are long. Primary production at the bottom of the food chain occurs almost exclusively among unicellular phytoplankton. Phytoplankton are eaten by other unicellular zooplankton or by marine copepods. The copepods therefore occupy the same trophic level as terrestrial grazers, such as rabbits or cows; in fact, copepods are on a higher trophic level because they do not shy away from eating unicellular zooplankton or a bit of cannibalism. Copepods form the base of fish production. Fish larvae and zooplanktivorous fish therefore occupy trophic level three or higher—they are the lions of the oceans.

The food chain length can be deduced from the observed predator:prey size ratio in the stomach. The ratio corresponds to the weight ratio between trophic levels. If the body size where energy enters the ecosystem (trophic level zero) is w_{zero}, then the trophic level v of predators of weight w is

$$w(v) = \beta_{PPMR}^{v} w_{zero} \Leftrightarrow v = \log\left(\frac{w}{w_{zero}}\right) \frac{1}{\log(\beta_{PPMR})}. \tag{2.23}$$

With the average predator:prey mass ratio of consumed prey $\beta_{PPMR} \approx 700$ (table 2.2) and a size of typical primary producers on the order of $w_{zero} \approx 10^{-9}$ g (Andersen et al., 2015), a 100 kg tuna is predicted to occupy trophic level ≈ 5. This prediction is in accordance with data from stable isotopes showing that marine food chains have between five and six trophic levels (Zanden and Fetzer, 2007).

2.7   WHAT IS THE TROPHIC EFFICIENCY?

Another derivation of the size spectrum exponent exists that differs from the one presented in section 2.4. This derivation was made by Kerr, Sheldon, and co-workers (Kerr, 1974; Sheldon et al., 1977) from the concept of a trophic transfer efficiency developed by Lindeman (1942).[4] By combining the Sheldon-Kerr derivation with the one in eq. 2.21, we can derive the value of the trophic efficiency.

Lindeman perceived the food chain as an engine transferring energy or biomass between trophic levels. The biomass transferred from one trophic level is the production of that trophic level. The efficiency of the trophic transfer is the ratio between the production received by a trophic level from the trophic level below and the production it delivers to the trophic level above. The production of the trophic level $i - 1$ is therefore the same as the total consumption by all individuals in the trophic level above, $\mathscr{C}_i$. The "trophic efficiency" ε_T can then be derived as the ratio of total consumption of all individuals in two consecutive trophic levels: $\varepsilon_T = \mathscr{C}_{i+1}/\mathscr{C}_i$. Following eq. 2.7, the total consumption of a trophic level i is $\mathscr{C}_i \propto w_i^n \mathscr{N}_i$, where $\mathscr{N}_i$ is the total number of individuals in the trophic level, and w_i is the average size of individuals in the trophic level. Inserting the consumption in the definition of the trophic level leads to $\varepsilon_T = w_{i+1}^n \mathscr{N}_{i+1}/(w_i^n \mathscr{N}_i)$. Defining the biomass as $B_{\mathrm{prey}.i} = \mathscr{N}_i w_i$ and relating weights of individuals in two trophic levels by the predator:prey size ratio, $w_{i+1} = \beta_{\mathrm{PPMR}} w_i$ gives

$$\frac{B_{\mathrm{prey}.i+1}}{B_{\mathrm{prey}.i}} = \varepsilon_T \beta_{\mathrm{PPMR}}^{1-n}. \tag{2.24}$$

The preceding result can be related to the size spectrum exponent by noting that the biomass in a trophic level is the same as the biomass of prey—that is, $B_{\mathrm{prey}.i} = B_{\mathrm{prey}}(w_i)$—defined in eq. 2.19. Inserting eq. 2.19 into eq. 2.24 and reducing leads to the trophic efficiency being expressed in terms of the exponent of the community size spectrum λ and the predator:prey mass ratio β_{PPMR} (Borgmann, 1987; Andersen et al., 2009b):

$$\varepsilon_T = \beta_{\mathrm{PPMR}}^{1+n-\lambda} = \beta_{\mathrm{PPMR}}^{2n-1-q}, \tag{2.25}$$

where the result from eq. 2.21, $\lambda = 2 + q - n$, has been used for the last equality. Using the values in table 2.2 leads to $\varepsilon_T \approx 0.14$. The trophic efficiency quantifies

---

[4] Actually, two more derivations of the size spectrum exponent have been proposed, both based on considerations of the predator-prey interactions. One is similar to the explanation in section 2.4 but does not constrain the metabolism of individuals with the "metabolic" law (eq. 2.7) (Benoît and Rochet, 2004; Datta et al., 2010). Therefore, depending on the parameters, a consumption rate of individuals derived from that explanation typically scales differently than the observations in fig. 2.5. This derivation is therefore not very satisfying. The other derivation is based on the metabolic theory of ecology, and it does indeed rely on the scaling of consumption with exponent 3/4 (Brown et al., 2004). Unfortunately, it also relies on the concept of "energy equivalence" (Damuth, 1987), which is an empirical relation established from observed size distributions of organisms. Since the derivation then implicitly uses the result it derives, it is based on a tautology. Fortunately, the mathematical result turns out to be equivalent to Sheldons (eq. 2.24), so the harm done is only aesthetic.

the loss of energy in the trophic transfer, and with a value of 14 percent aquatic ecosystems are more efficient than typical terretrial ecosystems with efficiencies in about 10 percent or less. The relation eq. 2.25 provides a way to calculate the trophic efficiency used in classic food-chain arguments, such as those developed by Sheldon and Kerr, on the basis of the physiology of individuals.

2.8   SUMMARY

The size spectrum theory developed in this chapter describes mass flows in marine ecosystems. The theory itself is based on the relations between physiology and body size—clearance rate (eq. 2.6), consumption rate (eq. 2.7), and prey size preference (eq. 2.8)—which together define the interactions between predators and prey. These individual-level rates form the basis of the derivation of the size spectrum (eq. 2.21) as a power-law function. From the size spectrum follows the predation mortality (eq. 2.22) as a decreasing function of body size. The theory adheres to the general philosophy in size spectrum theory of scaling from individual-level rates to a higher level or organization, here the entire ecosystem. This philosophy will later be used to scale toward a population (chapter 4) and the fish community (chapter 11). The basic power-law relations and the derived quantities are summarized in table 2.2.

Despite taking a somewhat cavalier approach to statistics, I insist on establishing a direct relation between data of the fundamental rates, and, as far as possible, a quantitative comparison with predictions. It is reassuring that the scaling and even the magnitude of the Sheldon spectrum and the observed mortality are so well predicted by the independently determined individual-level rates. While these correspondences should not be taken for a proof that eq. 2.18 is correct, this indicates that the assumptions outlined in this chapter are sound and that the parameter values are in the right ballpark. This is important, because it provides some confidence in the predictions in the remainder of this book.

CHAPTER THREE

Individual Growth and Reproduction

Developing a size spectrum theory for specific populations requires a more detailed description of the individual than I used in the previous chapter. In this chapter, I determine the growth and reproduction rates from the consumption rate, by developing an energy budget of the individual as a function of size. This chapter essentially seeks an answer to the question: how does an individual make use of the energy acquired from consumption?

Setting up energy budgets of individuals to determine growth rate is a tried and tested discipline, starting with Pütter in the early twentieth century, then developed into a robust tool by von Bertalanffy (1957) in his work on *Quantitative Laws in Growth and Metabolism*, and later extended to the type of biphasic growth model that I use (Ursin, 1979; Lester et al., 2004; Quince et al., 2008). Such energy budgets depend upon parameters that describe the individuals in the population. The simplest growth models, like the von Bertalanffy growth equation, rely on just two parameters, while the more complex models require more parameters. Clearly, the huge variety of life histories among fish—from small forage fish to large piscivores and from sluggish sunfish to highly active tuna—is better represented by a growth model with many rather than few parameters. However, from the perspective of building a simple theory, a large number of parameters makes it difficult to uncover simple relationships. I reduce the number of parameters by formulating the growth model using so-called life-history invariants, which are parameters that do not vary systematically between species. While the formulation of the growth model in terms of life-history invariants is largely successful, there is in particular one parameter that is not invariant between life histories: the *asymptotic size* (maximum size) of individuals in the population. This parameter plays the role of a master trait that characterizes most of the variation between life histories.

The energy budget accounts for all fluxes of mass and energy within the individual: assimilation losses, metabolic losses, as well as energy spent on reproduction and growth. The budget formalizes the decision made by the individual based on its state (hunger, size, maturation) and its life-history strategy. These decisions may be understood in a life-history optimization framework as those that optimize

---

BOX 3.1

LENGTH- AND WEIGHT-BASED VON BERTALANFFY MODEL

The weight-based von Bertalanffy growth model (eq. 3.4) can be transformed to a length-based representation by using the chain rule for differentials and the relation between weight and lenghts $w = cl^3$ (table 2.1)

$$\frac{dw}{dt} = \frac{dcl^3}{dt} = c\frac{dl^3}{dl}\frac{dl}{dt} = 3cl^2\frac{dl}{dt}. \tag{3.1}$$

Inserting into eq. 3.4 gives

$$\frac{dl}{dt} = \frac{1}{3}(c^{-1/3}A - kl). \tag{3.2}$$

The length-based model is commonly written in terms of the parameters $K = k/3$ and $L_\infty = q^{-1/3}A/k$. Inserting those definitions gives

$$\frac{dl}{dt} = K(L_\infty - l). \tag{3.3}$$

---

individual fitness. Fragments of such a theory have surfaced—for example, for the reproduction schedule (Charnov and Gillooly, 2004; Lester et al., 2004; Thygesen et al., 2005; Jørgensen and Fiksen, 2006; Quince et al., 2008)—but a complete theory has yet to emerge. Consequently, I will largely rely on empirical estimation of the life-history parameters in the growth model. The aims of this chapter are therefore (1) to develop descriptions of growth and reproduction of individuals; (2) to establish the asymptotic size as a master trait; and (3) to determine the parameters in the energy budget in terms of the individual size, the asymptotic size, and a set of life-history parameters.

3.1   THE VON BERTALANFFY GROWTH MODEL

The growth model I develop belongs to the family of biphasic growth models (Ursin, 1979; Lester et al., 2004; Quince et al., 2008). These models divide growth into juvenile and adult phases. Juveniles use all acquired energy for somatic growth, while adults divide energy between growth and reproduction. In this manner, the biphasic growth model accounts for the energy spent on reproduction, which is needed to derive the population size spectrum in the next chapter. Biphasic growth models owe their fundamental concepts to the von Bertalanffy growth model, and it is instructive to first look at that model.

The von Bertalanffy growth model describes growth rate dw/dt as the difference between two processes (von Bertalanffy, 1957):

$$\frac{dw}{dt} = Aw^n - kw. \tag{3.4}$$

The two processes represent acquisition of energy Aw^n and losses kw, or, in the words of von Bertalanffy, "anabolic" and "catabolic" processes. The coefficients A and k describe the overall level of the processes, while the exponents n and 1 describe how they scale with size. Regarding the exponent n, von Bertalanffy argued that acquisition was limited by anabolic processes—that is, those that involve absorbing oxygen or food across a surface (gills or the digestive system). Fish, he argued, are limited by the simple surface rule—that is, $n = 2/3$ (see p. 23). With that exponent, and the standard relation between length l and weight $w = cl^3$, eq. 3.4 can be rewritten in the common length-based form (box 3.1)

$$\frac{dl}{dt} = K(L_\infty - l), \tag{3.5}$$

with the solution

$$l(t) = L_\infty(1 - e^{-Kt}), \tag{3.6}$$

where L_∞ is the asymptotic length and K the von Bertalanffy growth constant with dimensions time^{-1}

$$K = \frac{A}{3c^{1/3}} \frac{1}{L_\infty} \quad \text{for } n = 2/3. \tag{3.7}$$

The resulting length-at-age curve in fig. 3.1a initially increases linearly with age with rate KL_∞. This increase follows from eq. 3.5, where $dl/dt \approx KL_\infty$ when $l \ll L_\infty$. Combining with eq. 3.7, we see that the initial growth rate in length is proportional to the growth coefficient A. As length approaches the asymptotic length L_∞, growth rate decreases until $dl/dt = 0$. The length-based von Bertalanffy growth equation has been very succesful in fisheries because it is simple, because it describes observed size-at-age curves fairly well, and because it is formulated according to body length that is easily measured. Consequently, almost all growth measurements of fish are reported via the two von Bertalanffy growth parameters K and L_∞.

Though popular, the mathematical form of the length-based von Bertalanffy equation is unfortunate from a statistical point of view, because the two parameters K and L_∞ are correlated (fig. 3.2 and eq. 3.7). Therefore, uncertainty in the estimation of one parameter will rub off on the other. When the growth function is fitted to data from commercially caught fish, there are usually only a few measurements

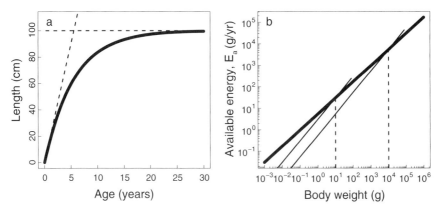

FIGURE 3.1. (a) von Bertalanffy length-at-age curve for a species with asymptotic length $L_\infty = 100$ cm and $K = 0.18$ yr^{-1} (eq. 3.6). The slanted dashed line is age multiplied by KL_∞, and the horizontal dashed line is at $l = L_\infty$. (b) Illustration of how the asymptotic size is determined by the available energy $Aw^{0.75}$ (thick line) and losses kw (thin lines), shown for two species with asymptotic sizes $W_\infty = 10$ g and 10 kg (dashed vertical lines).

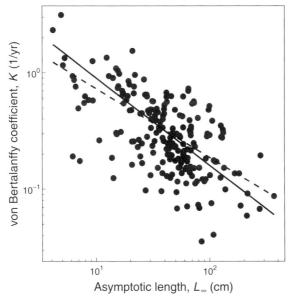

FIGURE 3.2. von Bertalanffy growth parameter K as a function of asymptotic length. The dashed line is a fit giving $K = CL_\infty^{-0.59}$ with $C = 2.85$ cm$^{0.59}$/yr; the solid line is a fit with fixed exponent -0.75, giving $C = 5.07$ cm$^{0.75}$/yr. Data points for teleosts from literature compilations (Kooijman, 2000; Gislason et al., 2010; Olsson and Gislason, 2016), are corrected to 15°C using a $Q_{10} = 1.83$.

of the largest individuals, simply because these are fished out of heavily exploited populations. In that case, the estimation of L_∞ becomes uncertain. Because of the correlation between K and L_∞, this uncertainty leads to an uncertainty in the estimation of K: if L_∞ is overestimated, K will be underestimated and vice versa. Therefore, the estimation of the initial growth rate KL_∞ becomes more uncertain than it needs to be, and possibly with systematic bias depending on how L_∞ is estimated. Had the relation eq. 3.7 been inserted into the von Bertalanffy growth equation (eq. 3.6) such that A and L_∞ were estimated instead, the uncertainty would have been confined to L_∞, whereas A would be reliably estimated from the data on juvenile fish.

The length-based von Bertalanffy growth equation (eq. 3.6) is based on $n = 2/3$. Von Bertalanffy should have read Haldanes (1928) essay "On Being the Right Size," with the gifted insight that the 2/3 law is too simple for describing how anabolic processes limit uptake (see p. 23). Perhaps West et al. (2001) read Haldane, because in their reformulation of the von Bertalanffy growth equation they used the 3/4 exponent to represent the fractal nature of uptake surfaces (West et al., 1997). Anyway, whether one value of the exponent is used over the other is not crucial—though there are indications that the 3/4 exponent leads to a better description of growth than 2/3 (Essington et al., 2001). Following in the footsteps of metabolic ecology, I will use $n = 3/4$ as the metabolic exponent.

3.2   ASYMPTOTIC SIZE AS A MASTER TRAIT

The von Bertalanffy size-at-age curve is shaped by the changing importance of the acquisition and loss terms, Aw^n and kw, as the individual ages and increases in size (fig. 3.1b). Because the two terms have different scaling exponents, n and 1, they will not be proportional to one another but losses will take an increasingly

---

BOX 3.2

ESTIMATING A FROM THE VON BERTALANFFY PARAMETERS K AND L_∞

A relation between the growth coefficient A and the length-based von Bertalanffy parameters K and L_∞ is provided in eq. 3.7 (Andersen et al., 2009a). This relation, however, is valid only for $n = 2/3$, while I use $n = 3/4$. We therefore need to develop a correction when we use $n = 3/4$.

Juvenile growth from the length-based von Bertalanffy equation can be estimated as the slope of the growth equation for $l \ll L_\infty$ as KL_∞. From the weight-based

---

(continued)

(Box 3.2 *continued*)

equation, juvenile growth is Aw^n. The length-based growth can be transformed to weight-based growth with the chain rule for differentials and the relation between weight and lenghts $l = (w/c)^{1/3}$ (table 2.1)

$$\frac{dl}{dt} = \frac{d(w/c)^{1/3}}{dt} = c^{-1/3}\frac{dw^{1/3}}{dw}\frac{dw}{dt} = \frac{c^{-1/3}}{3}w^{-2/3}\frac{dw}{dt}. \tag{3.8}$$

Inserting the juvenile growth from the von Bertalanffy parameters $dl/dt = KL_\infty$ and $dw/dt = Aw^n$ and rearranging gives

$$A = 3c^{1/3}w^{2/3-n}KL_\infty. \tag{3.9}$$

If we use $n = 2/3$, the dependency on weight disappears, and we recover the simple relation between K, L_∞, and A from eq. 3.7. For $n = 3/4$, there is an additional factor $w^{-1/12}$ to consider. I, rather arbitrarily, assert that growth rate of the two models should be equivalent at the size at maturation $w = w_m = \eta_m W_\infty$. This gives

$$A = 3c^{1/4}\eta_m^{-1/12}KL_\infty^{3/4} \approx 1.14KL_\infty^{3/4} \quad \text{for} \quad n = 3/4, \tag{3.10}$$

where the last approximation used $c \approx 0.01$ g/cm^3 (table 2.1) and is valid for K measured in years^{-1}, L_∞ in cm, and A in g$^{1/4}$/yr.

large share of the available energy as an organism grows in size, leaving less and less for growth. At some size, all available energy is used for losses and growth stops. This size defines the asymptotic length L_∞ or asymptotic weight W_∞ of individuals in the species. The asymptotic weight can be derived from eq. 3.4 as the size where growth stops ($dw/dt = 0$)

$$W_\infty = \left(\frac{A}{k}\right)^{1/(1-n)}. \tag{3.11}$$

This equation establishes a relation between the asymptotic weight W_∞, the coefficient of acquired energy A, and losses k. It is a key relation because it shows the existence of a *trade-off* between A and k that determines the asymptotic size: large species (large W_∞) either acquire more energy (higher A) or have smaller weight-specific losses (smaller k). In this way, the differences in growth between species are defined solely by the asymptotic size and the growth coefficient A. With the relation eq. 3.11, the von Bertalanffy growth model (eq. 3.4) can be rewritten as

$$\frac{dw}{dt} = Aw^n\left[1 - \left(\frac{w}{W_\infty}\right)^{1-n}\right]. \tag{3.12}$$

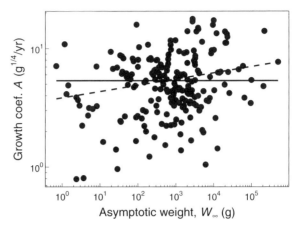

FIGURE 3.3. The growth coefficient A derived from eq. 3.10. The dashed line shows fit to a power law (exponent 0.053); the solid line is the geometric mean, $A = 5.35$ $g^{0.25}$/yr. Data from Kooijman (2000); Gislason et al. (2010); Olsson and Gislason (2016), corrected to 15°C.

The growth coefficient A represents processes related to energy acquisition and assimilation, and we can expect that these processes are unrelated to the asymptotic size of the species. To find the value of A, we have to rely on measurements of the von Bertalanffy growth parameters K and L_∞ from the length-based size-at-age curves. The simple relation between the growth coefficient A and the von Bertlanffy parameters in eq. 3.7 is only valid for $n = 2/3$, and not for $n = 3/4$; however, a decent approximation of A for $n = 3/4$ is derived in box 3.2. The compilation of growth data in fig. 3.3 shows how A is indeed roughly independent of asymptotic size, or perhaps slightly increasing. The data also reveal a substantial variation in growth rates between species with similar asymptotic size, by around a factor of 2 to either side of the mean. The slight increase of the growth coeffient with asymptotic size is often reported in other studies (Pope et al., 2006; Olsson and Gislason, 2016). It suggests that larger species tend to have faster growth than smaller species. Faster growth would be accompanied by elevated body temperature and higher metabolic rates, which is indeed found among the scombroids (tunas, swordfish, which), and so on are a dominant group of larger fish species (Killen et al., 2016). For simplicity, I will use A as a constant in the following.

Biphasic Growth Equation

The biphasic growth model is modeled on top of the skeleton provided by von Bertalanffy. The main differences is that life is divided into juvenile and adult stages: in the juvenile stage, all assimilated energy is used for growth, while adults

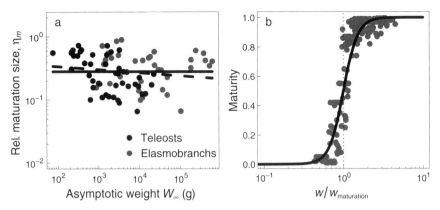

FIGURE 3.4. Cross-species analysis of size at maturation. (a) Size at maturation relative to asymptotic size, η_m. Power-law fits to all data (dashed line), and with exponent 0 (solid line). The average value is 0.28. Data from Olsson and Gislason (2016). (b) Average maturation of North Sea saithe fitted to the maturation function eq. 3.15. The fit gave a steepness of the function $u \approx 5$ (data from ICES stock assessment).

also use energy for reproduction. We can write juvenile growth rate as

$$g_j(w) = Aw^n \quad \text{for juveniles.} \qquad (3.13)$$

Individuals mature at a size w_m. Size of maturation is roughly proportional to asymptotic size $w_m = \eta_m W_\infty$, where the constant of proportionality is $\eta_m \approx 0.28$ (fig. 3.4). Mature individuals invest some fraction of their acquired energy into reproduction, typically proportional to their weight. As the investment into reproduction scales linearly with weight, it belongs to the loss term in the von Bertalanffy equation, but only for adults. Adult growth then become:

$$g(w) = Aw^n - kw \quad \text{for adults, } w > \eta_m W_\infty. \qquad (3.14)$$

Juvenile and adult growth are brought together by a maturation function $\psi_m(w/w_m)$, which switches smoothly between zero and 1 when the argument $w/w_m = 1$ at the size of maturation

$$\psi_m(w/w_m) = [1 + (w/w_m)^{-u}]^{-1} = [1 + (w/(\eta_m W_\infty))^{-u}]^{-1}, \qquad (3.15)$$

where the exponent $u \approx 5$ determines the steepness of the function (see fig. 3.4b). Introducing this function, the combined growth equation is

$$g_{\text{bp}}(w) = Aw^n - \psi_m(w/w_m)kw, \qquad (3.16)$$

where I use the subscript bp to signify the biphasic growth equation. Just as with the von Bertalanffy model, the biphasic growth model in eq. 3.16 can be formulated in terms of asymptotic size. Turning the relation between A, k,

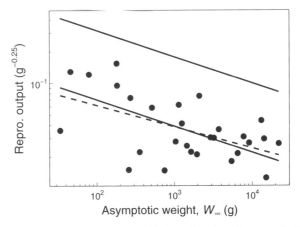

FIGURE 3.5. Specific reproductive output $R_{egg}/(Aw)$ as a function of asymptotic size. The reproductive output is shown as the annual egg production (measured in weight per year) divided by the growth constant A and by weight. The upper solid line is the value of $k/A = W_\infty^{n-1}$. The value of A for each species is calculated from the age at maturation with eq. 3.25. The dashed line is a fit giving exponent -0.20; the lower solid line is fit with exponent fixed to -0.25. The value of ε_{egg} is estimated as $\varepsilon_{egg} \approx 0.22$. Data from Gunderson (1997).

and asymptotic size (eq. 3.11) around reveals how total losses, and thereby the reproductive investment, is related to asymptotic size

$$k = AW_\infty^{n-1}. \tag{3.17}$$

Inserting eq. 3.17 back into the growth model eq. 3.16 gives a trait-based formulation of the biphasic growth model

$$g_{bp}(w) = Aw^n \left[1 - \psi_m \left(\frac{w}{\eta_m W_\infty} \right) \left(\frac{w}{W_\infty} \right)^{1-n} \right]. \tag{3.18}$$

The biphasic growth model does not allow an analytical solution for weight-at-age, but analytical solutions to juvenile growth are given in box 3.3. Fig. 3.6 shows numerical solutions to size-at-age from $dw(t)/dt = g_{bp}(w)$.

The main advantage of the biphasic growth model over the von Bertalanffy model is that it accounts for investment in reproduction (eq. 3.17) and shows how reproduction scales with asymptotic size. The exponent of the investment in reproduction is negative (-0.25), and the investment in reproduction per weight k is therefore decreasing with asymptotic size. This decreasing pattern is also observed in data on the reproductive output (fig. 3.5) (Gunderson, 1997; Charnov et al., 2001; Olsson and Gislason, 2016). The reproductive investment kw is not only spent on producing eggs, it also represents other aspects of reproduction such as a spawning migration. Assuming that the additional energy used is proportional to the mass of eggs produced per time, the individual reproductive output R_{egg} (mass of eggs per time per individual) becomes

$$R_{egg}(w) = \varepsilon_{egg} k w = \varepsilon_{egg} A W_\infty^{n-1} w. \tag{3.19}$$

The value of "reproductive efficiency" is estimated to be $\varepsilon_{egg} \approx 0.22$ (fig. 3.5).

BOX 3.3

ANALYTICAL SOLUTIONS OF SIZE-AT-AGE

Von Bertalanffy (1957) gave the solution of the von Bertalanffy growth model (eq. 3.4) as

$$w(t) = \left[\frac{A}{k} - \left(\frac{A}{k} - w(0)^{1-n} \right) e^{-(1-n)kt} \right]^{1/(1-n)}, \tag{3.20}$$

where $w(0)$ is size at age zero. For the value of the catabolic coefficient k, we can use the relation with asymptotic size from eq. 3.11 $k = AW_\infty^{n-1}$ to give

$$w(t) = \left[W_\infty^{1-n} - \left(W_\infty^{1-n} - w(0)^{1-n} \right) e^{-(1-n)AW_\infty^{n-1}t} \right]^{1/(1-n)}. \tag{3.21}$$

If the asymptotic size is much larger than offspring size, $w(t)$ is well approximated by

$$w(t) = W_\infty \left(1 - e^{-(1-n)AW_\infty^{n-1}t} \right)^{1/(1-n)} \quad \text{for} \quad W_\infty \gg w(0). \tag{3.22}$$

Age at maturation can be approximated from eq. 3.22 as the age where the weight is size at maturation—that is, where $w(t) = \eta_m W_\infty$. This gives

$$t_{mat} \approx W_\infty^{1-n} \frac{\ln(1 - \eta_m^{1-n})}{A(n-1)} \approx 0.75 W_\infty^{1-n} \quad \text{for} \quad W_\infty \gg w(0), \tag{3.23}$$

where the approximation is valid if weight is measured in units of grams and time in years. The age at maturation scales "metabolically" with asymptotic size to the power $1 - n \approx 0.25$ and inversely with the growth coefficient A; a slower growth rate leads to a late age at maturation.

The biphasic growth equation can be solved for juvenile growth

$$w(t) = \left(A(1-n)t + w(0)^{1-n} \right)^{1/(1-n)} \approx (A(1-n)t)^{1/(1-n)}, \tag{3.24}$$

with age at maturation being

$$t_{mat} \approx \frac{\eta_m^{1-n}}{A(1-n)} W_\infty^{1-n} \approx 0.67 W_\infty^{1-n}, \tag{3.25}$$

again with the approximation being valid when W_∞ is measured in grams and age in years.

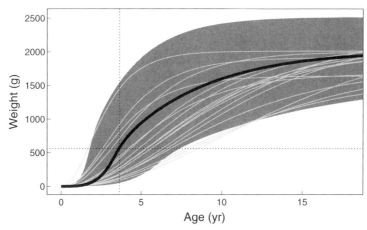

FIGURE 3.6. Growth curve for a species with $W_\infty = 2$ kg found by solving eq. 3.18 (thick black line). The gray lines are von Bertalanffy growth curves calculated with the observed parameters from fig. 3.3 from species with asymptotic sizes in the range 1.6 and 2.5 kg, and the gray patch is the solution to eq. 3.18 with asymptotic sizes in the same range and A varying with a coefficient of variation of 1.95. This illustrates how variation in growth between species with similar asymptotic size is roughly a factor of 2. The dotted lines show size at maturation as $\eta_m W_\infty$ and age at maturation approximated with eq. 3.25.

The trait-based growth model in eq. 3.18 is formulated in terms of parameters that are expected to be roughly invariant between species, A, ε_{egg}, η_m and n, and with the asymptotic size as the main trait that characterizes growth and reproduction of a species. Formulating growth with a trait-based model makes it possible to make general statements about the differences between small and large species just by varying W_∞. Of course, if some additional information about a specific species is available, such as the growth coefficient A, then this information should be used to described the species' growth more accurately. The growth equation can therefore be used equally well as trait-based description of growth, with W_∞ being the trait and all other parameters constant, or as a model of a specific species with all parameter values being specific to that species.

3.3 BIOENERGETIC FORMULATION OF THE GROWTH EQUATION

The biphasic growth equation does the job it was given: it describes growth and reproduction as a function of size. That in itself is sufficient for the single-species calculations in parts II and III. However, it is insufficient for a dynamic description of growth needed in part IV. Further, the central parameter, the growth

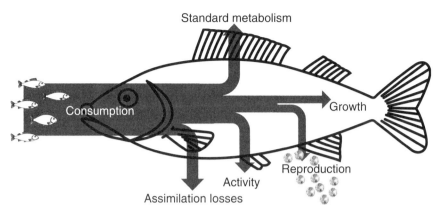

FIGURE 3.7. Sketch of the energy budget. Consumed food is lost due to inefficient assimilation and energy needed for assimilation. The assimilated energy is used to fuel standard metabolism and activity. The remaining available energy is divided between growth and reproduction.

coefficient A, was determined only from empirical data. How is the growth coefficient connected to the fundamental physiological assumptions developed in chapter 2? To answer this question, I will dig deeper into the metabolic processes by considering a complete energy budget of an individual.

The biphasic growth model developed is based on von Bertalanffy's idea that processes can be divided into two parts: *anabolic* processes related to acquisition of energy (Aw^n) and *catabolic* processes associated with losses (kw). This led von Bertalanffy to conclude that respiration was associated with the catabolic kw processes. That interpretation of growth and metabolism in fish is a simplification: losses also occur during the acquisition of energy, notably during assimilation. Further, some of the losses in the kw term are not associated with metabolism but with the reproductive output. The development of an energy budget will clarify exactly where losses are occurring.

An energy budget states how consumption $C(w)$ is used to fuel the processes of assimilation M_{assim}, standard metabolism M_{std}, activity M_{act}, reproduction $R_{egg}(w)/\varepsilon_{egg}$, and growth $g(w)$ (fig. 3.7)

$$C(w) = M_{assim}(w) + M_{std}(w) + M_{act}(w) + R_{egg}(w)/\varepsilon_{egg} + g(w). \qquad (3.26)$$

All terms are mass rates with units of wet weight per time. Wet weight is strictly speaking not an energy, so how can this be an energy budget? The implicit assumption that allows equating mass with energy is that wet weight is proportional to energy (1 g of wet weight equals roughly 5.5 kJ), so if eq. 3.26 is divided by 5.5 g/kJ on both sides, it becomes an explicit energy budget. However, in the remainder it is not necessary to distinguish between wet weight and energy, so this conversion is ignored.

Consumption and Assimilation

In chapter 2, consumption rate was described as $C(w) = f_0 h w^n$ (eq. 2.18), where the *feeding level* f_0 is the consumption as a fraction of maximum consumption hw^n. Not all consumed mass and energy is assimilated; some is lost during the assimilation process, M_{assim}, owing to incomplete uptake (egestion and excretion), and owing to the metabolic expenditure of the uptake (the *specific dynamic action*). All of these processes can be taken to be proportional to the consumption. Kitchell et al. (1977) estimated the specific dynamic action to be 15 percent of food consumption and conservative estimates of egestion and excretion to be 15 percent and 10 percent, respectively. This results in assimilation losses $M_{assim} = (1 - \varepsilon_a) f_0 h w^n$, with the assimilation efficiency being $1 - 0.15 - 0.15 - 0.1$—that is, $\varepsilon_a = 0.6$. The assimilated consumption is then

$$C_{assim}(w) = \varepsilon_a f_0 h w^n. \tag{3.27}$$

Standard Metabolism and Activity

The assimilated consumption partially describes the acquisition term Aw^n in the von Bertalanffy equation (eq. 3.4); however, there are also further losses to standard metabolism and activity. The standard metabolism is the energy required to maintain the basic metabolic processes that keep the organism alive, while activity is the energy spent on foraging, migration, and so on. The standard metabolism of fish was described in fig. 3.5 as being $M_{std}(w) = k_s w^n$, where again n is the metabolic exponent. Activity metabolism is difficult to measure because it depends on the level and frequency of activity. In the absence of information, I will assume that it is simply proportional standard metabolism and represented within the coefficient k_s. The standard and activity metabolism is proportional to maximum consumption rate hw^n. To reflect this relation, metabolism is represented as a fraction of maximum assimilated consumption, $k_s = f_c \varepsilon_a h$, where f_c is the *critical feeding level* —that is, the fraction of maximum assimilated consumption rate required to stay alive and feed

$$M_{std}(w) + M_{act} = f_c \varepsilon_a h w^n. \tag{3.28}$$

A reasonable value of f_c is around 0.2 (Hartvig et al., 2011).

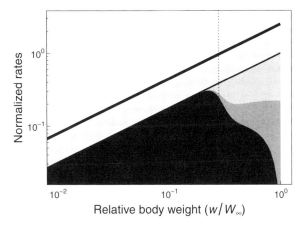

FIGURE 3.8. Illustration of the energy budget in eq. 3.30 as a function of size relative to asymptotic size with the vertical line showing size at maturation. The lines show (from the top and down): consumption rate $f_0 h w^n$ (thick line); excretion and egestion (white area); standard metabolic losses and specific dynamic action (digestion, light gray); available energy $A w^n$ (thin black line); metabolic losses associated with reproduction (somewhat darker gray area); reproductive output (even darker gray); growth (very dark gray). All rates are scaled with the available energy at the asymptotic size $A W_\infty^n$, and the division of energy is therefore independent of asymptotic size. Note that because the y-axis is logarithmic, the area of a patch is not proportional to the absolute amount of energy.

The energy budget defined in eq. 3.26 can now be assembled and the growth rate $g(w)$ isolated on the left-hand side

$$g(w) = \varepsilon_a f_0 h w^n - f_c \varepsilon_a h w^n - \psi_m(w/w_m) k w. \qquad (3.29)$$

Collecting terms according to whether they are proportional to w^n or w gives

$$g(w) = \underbrace{\varepsilon_a h(f_0 - f_c)}_{A} w^n - \psi_m(w/w_m) k w. \qquad (3.30)$$

This is the same as the biphasic growth eq. 3.16, if we define the growth coefficient as

$$A = \varepsilon_a h(f_0 - f_c). \qquad (3.31)$$

Fig. 3.8 illustrates how energy is divided between the different processes as a function of size.

The relation between A and maximum consumption h was used in chapter 2 to estimate the maximum consumption used in fig. 3.5. The physiological parameters that make up A—the assimilation efficiency ε_a, the coefficient of maximum consumption h, and the critical feeding level (standard metabolism) f_c—are constants that are expected to vary little between species. The feeding level f_0 will also on average be around 0.6. In part IV, however, I will consider dynamic models, and there the feeding level will no longer be constant but will depend on available food. For now, however, the feeding level is considered constant, so all these parameters can be combined into a constant growth coefficient.

3.4   WHICH OTHER TRAITS DESCRIBE FISH LIFE HISTORIES?

I have used the asymptotic size W_∞ as the master trait that determines the variable processes in the growth model. But are these processes really determined by W_∞? For instance, eq. 3.11 related W_∞ to the growth coefficient A and the investment in reproduction k as $W_\infty = (A/k)^4$. That relation could just as well be used to state that the decision about the reproductive investment k together with A determines W_∞. That is correct, so why not choose the reproductive investment k as the master trait and let that determine W_∞? The choice of W_∞ is one of convenience: it is simply the easiest trait to determine in a population of fish. Even for the most data-limited populations, we have a reasonable guess of the asymptotic size, simply as the largest fish observed. Using W_∞ as the master trait means that it will be possible to apply the theory to even very data poor stocks. Further, W_∞ has a more intuitive meaning than k. Knowing for example that $W_\infty = 10$ kg, I can immediately state that the stock in question is probably piscivorous, while that would be a harder statement to make on the basis of knowing that $k = 0.44$ yr^{-1}.

The model of individual growth and reproduction suggests that many important life-history parameters correlate with asymptotic size W_∞ and/or the growth rate coefficient A: age at maturation, size at maturation, reproductive investment, and life span. This implies that knowing only asymptotic size W_∞ and the growth rate coefficient A, all other parameters can be estimated. While the relations between the parameters are borne out clearly in the theory, they are less clear in the data analyses shown in this chapter. How can alleged life-history "invariants," such as the ratio between asymptotic size and size at maturation, be considered invariants when they vary considerably between species? There are three aspects of an answer to this question: semantic, empiric, and theoretic.

Calling a measure, such as the size of maturation relative to the asymptotic size, an invariant does not imply an absence of variation between species. Rather, it implies that the parameter does not co-vary with other traits, such as the growth rate, or with the environment—for example, temperature—or with phylogeny— that is, that related species have similar deviations from the mean. Using the term *life-history invariant* is therefore not a statement that life histories do not vary between species, but rather that the variation is random and unexplained, at least according to our current level of understanding. To avoid implying that these parameters are fundamentally invariant, I refer to them as *life-history parameters*, and not as life-history invariants.

Regarding the empiric basis of the life-history parameters, it should be remembered that the quality of data is limited. For instance, some of the data are not direct measurements. An example is the growth coefficient A, which is only indirectly estimated from measured growth curves; this clearly generates extra noise and possibly even systematic biases, as discussed on p. 40. Further, measured growth curves are rarely growth curves of individuals, but are based on the size-at-age of average surviving individuals. If faster growth is correlated with higher mortality, then the size-at-age curves are biased toward having more slow growers in the higher age classes than faster growers. Other traits, such as reproductive investments, are very difficult to estimate reliably: gonad weight may be a proxy of reproductive investment, but the annual reproductive investment requires knowledge of the average number of spawning events during a year, how spawning varies with food concentration, the degree to which skipped spawning occur, and the fraction of the investment used for migration and spawning behavior. Nevertheless, while all these considerations are relevant, they are insufficient to explain the variation of the data around the theoretically predicted relationships.

There must therefore be other life-history correlates that explain some of the variation that we observe, we just have not yet uncovered them. Take the variation in the growth rate coefficient in fig. 3.3 as an example: there is almost a factor of 10 difference between the slowest and the fastest growing fish, even when correcting for temperature. It seems like a safe bet that such profound differences in growth rates are correlated with other traits. Even though we do not know which traits, we can form hypotheses. One obvious candidate for another "trait axis" would be a slow-fast life history continuum. Slower growing individuals are assumed to obtain less food, and have a correspondingly lower clearance rate. A lower clearance rate would imply a lower activity and thus a lower critical feeding level f_c. However, clearance rates are difficult to measure directly and the activity coefficient is, in the words of my colleague and expert fish physiologist Niels Gerner Andersen, the "dark horse" of energy budgets—we know very little about in situ rates of activity. One way out of this problem is to assume that activity correlates with swimming speed, but then again, the fraction of time an individual is active versus passive also plays in. The difficulties of establishing credible data for the underlying processes for many species is a key reason why the trade-offs behind an obvious candidate for an additional trait like the slow-fast continuum have not been revealed.

Another, possibly related, trait is investment in defense. Fish live in an unforgiving environment where foraging implies an exposure to being eaten. The risk of being eaten can be lowered by investing in defense. Defenses can be manifested as spines (sticklebacks or perches), by being cryptic (sculpins), or by hiding (many flatfishes or sand eel). All these strategies have costs in terms of creating the defense and in the defensive behavior itself. Hiding, for instance, comes at a cost in forgone feeding. In other words, defense implies slower growth. Defense traits can therefore easily be confounded with the "slow" end of a the fast-slow life-history continuum. Quantifying the cost and benefits of defense traits, the trade-offs, is even more difficult than for the fast-slow continuum because they require answers to tricky questions: How much feeding does a hiding individual forgo? How much reduced mortality does this imply? This quantification has been done for small animals such as copepods (Kiørboe, 2011) that can be studied easily in the laboratory, but not for fish. In the following chapter, I will include a linear trade-off between investments in defense and mortality: investment in defense comes at a cost of lower A, which translates into lower mortality.

If the trade-offs related to the slow-fast and defense axes could be quantified, some part of the unexplained variation in for example, the growth rate in fig. 3.3, will be explained by the extra dimensions added to the trait-space. The variation might not go away entirely, but the clever addition of even more traits might lower it. The addition of more traits complicates the analysis of models and should therefore be done only if they add significant insights. The art in formulating trait-based descriptions lies in an inspired choice of the smallest set of traits that describe the largest amount of the observed variation in life-history parameters. I will revisit this aspect in chapter 9; the minimal model developed here uses just one trait: the asymptotic size.

3.5   SUMMARY

This chapter developed a simple von Bertalanffy–like biphasic model of growth and reproduction in fish. The model describes how growth and reproduction vary between fish species with different asymptotic sizes: species with large asymptotic size are expected to have a smaller reproductive output per body weight than smaller species. Other aspects of growth are determined by a set of life-history parameters. While these parameters are not exactly invariant between species, there is no systematic variation with asymptotic size. This establishes asymptotic size as the "master trait" to describe a fish stock.

The growth model predicts size-at-age curves that deviate from classical von Bertalanffy predictions. Notably, juvenile sizes-at-age are smaller than von Bertalanffy predicts. The deviations are unfortunate because the von Bertalanffy equation is generally perceived to be a fine representation of growth data (von Bertalanffy, 1957; Ursin, 1967). The reason for the deviations stem from me using $n = 3/4$ and not $n = 2/3$ for consumption and for losses to standard metabolism and activity. These choices are not in perfect accordance with the current knowledge of how metabolism scales with size. There are few measurements of how standard metabolism scales with weight as individuals grow in size, but it does seems to scale with an exponent higher than $3/4$; values between 0.8 and 0.9 are commonly reported (Killen et al., 2007). Losses to activity are harder to measure, but theoretical considerations indicate that they also scale with a higher exponent (Ware, 1978). Increasing the exponent of metabolism would introduce an extra term in juvenile growth (eq. 3.13) and bring it closer in form to the von Bertalanffy growth equation. An improved model would therefore include an explicit loss term in the growth equation with a different exponent from the acquisition, or, if the constraint of an equation with just two exponents (n and 1) were to be maintained, to include standard metabolic losses in the catabolic term.

There is one important reason for not including activity or other losses with an exponent higher than 3/4: loss rates scaling with a higher exponent than acquisition of energy—that is, higher than n, limits that maximum asymptotic size of all fish stocks. This limitation is illustrated in fig. 3.1b: higher asymptotic sizes means a lower coefficient of the rate in question. If the overall maximum size of any fish species is 1 ton, then the coefficient that scales directly with mass cannot be larger than $A(1 \text{ ton})^{-0.25} \approx 0.17 \text{ yr}^{-1}$ (found using 3.17). Such a term would be insignificant for small fish species in comparison to the reproductive investment, which for a 10 kg fish is 0.54 yr^{-1} and 3.0 yr^{-1} for a 10 g fish. Alternatively, we could assume that this constant term also changes with asymptotic size, such that it is smaller for large species (Andersen and Brander, 2009; Calduch-Verdiell et al., 2014; Andersen et al., 2015, 2016). There are, however, no observations about how the level of standard metabolism or activity varies with asymptotic size—that is, do larger species have lower levels of standard metabolism and activity than smaller species? There is no a priori reason to expect that the levels should decline with asymptotic size. Why should a large species be less active or have lower size-corrected standard metabolic levels than a small species? On the contrary, it is prudent to assume that the metabolic rates correlate with other traits, as argued in section 3.4. If a size-based description is developed for a particular stock, the maintenance of the dependency with asymptotic size is less of a concern. In that case, a higher degree of accuracy is desirable, which requires a more complex growth equation where rates vary more accurately with size, as for example, in Ursin (1967, 1979).

The choices made in the growth equation are in line with my willingness to sacrifice some accuracy in the interest of developing a coherent and general theory. This sacrifice makes it possible to formulate growth and reproduction in terms of asymptotic size and thereby make general statements about fish stocks broadly just by varying asymptotic size. Dividends on the sacrifice will be paid out in the following chapters, where the trait-based formulation of growth and reproduction reveals how density dependence and the impact of fishing vary across fish species broadly. In conclusion, the model of individual growth presented in eqs. 3.18 and 3.30 is far from perfect. It does, however, represent the best attempt at formulating a simple size-based growth equation where all the variation is represented by the variation in just one parameter: the asymptotic size W_∞.

Part II
POPULATIONS

CHAPTER FOUR

Demography

The first part of this book established the size structure of the entire ecosystem. The description rested upon the rule that bigger individuals eat smaller individuals and upon the limitations that organism size and physiology place on metabolic processes. From the perspective of an individual organism, the size structure of the ecosystem determines two central conditions for life: the availability of food and the predation risk from larger individuals. However, ecosystems and communities are not made up only of individuals, they are also made up of populations. In this second part, I combine the description of individual growth and reproduction (chapter 3) with the description of the ecosystem from the previous chapter to determine the structure of a population within the ecosystem (this chapter). The perspective of demography is mainly static, as it emerges when the food and predation environment are fixed. The dynamic interplay between the population itself and its food and predation is taken up in part IV. The static perspective on demography is the backbone of current fisheries advice and management, and I explore the consequences of fishing on the population (chapters 5 and 6). Some aspects of dynamics, though still in fixed food and predation environment, are considered in chapter 7.

The demographic structure of a population is described by a population size spectrum $N(w)$. The population size spectrum differs from the community size spectrum $N_c(w)$ from chapter 2, in that it represents the abundance of individuals in only one population. The range of sizes is therefore limited to the span from offspring size to the asymptotic size. In this chapter, I will show how the population size spectrum can be calculated if we know the vital rates of the individuals within the population: the growth rate, the reproduction rate, and the mortality as functions of size.

We can calculate the population size spectrum from our established knowledge of the mortality risk from chapter 2 and the growth rate from chapter 3. Both quantities are determined by the community: mortality stems from predation by larger individuals and growth from the abundance of prey. The explicit links between mortality and growth and the community size spectrum were established in chapter 2 and section 3.3. Here we will mostly leave those links aside and consider mortality and growth as given. As in chapter 3, differences between species are characterized mainly by the master trait, the asymptotic size W_∞.

At the core of size-based theory is metabolic scaling: growth scales with body size $\propto w^n$, and mortality $\propto w^{n-1}$, with n being the metabolic exponent. On this basis, we are led to the reasonable hypothesis that also population-level measures should obey metabolic scaling rules (Fenchel, 1974; Brown et al., 2004) (see box 4.1). For example, the growth rate of a population, being a rate, should scale with asymptotic size as W_∞^{n-1} (Economo et al., 2005; Andersen et al., 2009a). As $n \approx 0.75$, the scaling exponent is -0.25, meaning that the smaller species are expected to have a faster population growth rate than large species. For example, the population of a species with asymptotic size of 10 g would increase faster by a factor of $(10/10,000)^{-0.25} \approx 5.6$ than a species with asymptotic size 10 kg. I will show that this expectation does not hold: large species in general have higher reproductive rates and higher productivity than metabolic scaling rules predict.

BOX 4.1

METABOLIC SCALING RULES

Metabolic theory was created by an interdisciplinary group of scientists in the late 1990s to early 2000s at the Sante Fe Institute. The theory is based upon the observation that the resting metabolism of an individual scales with body mass roughly $\propto w^{3/4}$ (Kleiber's law). This scaling law was explained as the result of an optimal fractal delivery network (West et al., 1997). Later temperature scaling was added. The metabolic theory provided predictions about metabolic scaling rules (Brown et al., 2004), the von Bertalanffy growth equation (West et al., 2001), and scaling of size spectra (Brown et al., 2004), though the derivation of the community size spectrum is flawed (see note 4, p. 36).

The metabolic scaling rules are essentially derived by dimensional arguments. Since metabolism has dimensions mass/time, then all physiological rates, with dimensions 1/time, should be proportional to metabolism divided by body weight— that is, $\propto w^{3/4}/w = w^{-1/4}$. This argument applies to rates such as population growth rate, mortality rates, heart rates, developmental rates, even evolutionary rates.

The central role of metabolic scaling was well known before metabolic theory was explictly formulated—the exponent n was baptized the "metabolic exponent" already by Sheldon et al. (1977), and the metabolic scaling rule for population growth was given by Fenchel (1974). Nevertheless, the formulation of a coherent framework has made it possible to apply it to explain ecological patterns on broad scales, and metabolic theory is now a cornerstone of modern macro ecology.

This is a central insight that will help explain in the next chapter why fish are so remarkably resilient toward fishing. As fish demography does not obey metabolic scaling rules, the importance of the metabolic exponent n diminishes. The central role of n is superseded by a new parameter, the *physiological mortality*, that combines the essence of growth and mortality. I will discuss the physiological mortality at length, and try to estimate its value from meta-analyses of mortality and growth rate in fish.

The road ahead starts with solutions of the population size spectrum. I will follow two parallel tracks: a simplified analytical solution, and a full numerical solution. The analytical solution offers insights into the governing scaling relationships between asymptotic size and population-level measures, such as spawning stock biomass, reproductive output, and lifetime reproductive output, while the numerical solution allows us to explore the effects of size-based fishing in the next chapter. The population dynamical loop is closed with a consideration of how density-dependent recruitment is calculated from the reproductive output through a *stock-recruitment relation*.

4.1   WHAT IS THE SIZE STRUCTURE OF A POPULATION?

Classic fish demography, as synthesized by Beverton and Holt (1957), is age-based. It is formulated with life tables set up in matrices with each column representing survival of a cohort (age group). Solving the life tables is straightforward: the abundance or biomass in one cohort is the same as the year before multiplied by the survival. The formulation as life tables made numerical solutions possible before the proliferation of electronic computers. Fig. 1.1 shows Sidney Holt performing such numerical calculations with a Brunsviga mechanical calculator. Those calculations were used to build the cardboard model of the maximum yield from a population, seen in the background of the image.

To solve a size-structured population, we must factor the growth rate in, because it sets the speed by which individuals move from one size class to the next. The abundance of a size group is a balance between how many individuals grow into the group, how many grow out of it, and how many are dying (fig. 4.1). In a continuous-size spectrum, this balance is formalized in the McKendrick–von Foerster partial differential equation (box 4.2)

$$\frac{\partial N(w)}{\partial t} + \frac{\partial g(w)N(w)}{\partial w} = -\mu(w)N(w), \tag{4.1}$$

where $g(w)$ is the growth rate (mass per time) of individuals with weight w, and $\mu(w)$ is the mortality rate (per time). $N(w)$ is the population size spectrum

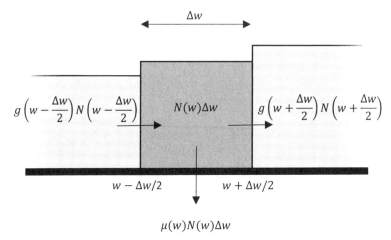

$$\mu(w)N(w)\Delta w$$

FIGURE 4.1. Sketch of the balance between growth and mortality in a size group containing $N\Delta w$ individuals in the size range $w \pm \Delta w/2$. The flux of individuals growing into the group (numbers per time) is the growth rate multiplied by the abundance at the boundary $g(w - \Delta w/2)N$ $(w - \Delta w/2)$, and likewise the flux of individuals growing out is $g(w + \Delta w/2)N(w + \Delta w/2)$. The flux of individuals disappearing from the group is the mortality multiplied by the abundance $\mu(w)N(w)\Delta w$.

BOX 4.2

THE MCKENDRIC–VON FOERSTER EQUATION

The McKendric–von Foerster equation can be derived by considering the losses and gains of individuals in a range of body sizes $w - \Delta w/2$ to $w + \Delta w/2$ (see fig. 4.1). We assume a size distribution $N(w)$, which represents the abundance density— that is, number of individuals per weight (see box 2.1). The number of individuals within that size range is $\approx N(w)\Delta w$. The rate at which individuals from the size class of smaller individuals grows into the range, the "flux" of individuals, is $g(w - \Delta w/2)N(w - \Delta w/2)$. During a short time interval, Δt, the number of individuals growing into the size range is then $g(w - \Delta w/2)N(w - \Delta w/2)\Delta t$. Similarly, the number of individuals growing out is $g(w + \Delta w/2)N(w + \Delta w/2)\Delta t$. In the same time interval, some individuals within the size range will die. The number of deceased individuals is approximated by the total number in the interval $N(w)\Delta w$ multiplied by the mortality $\mu(w)$ and the length of the time interval: $N(w)\Delta w\mu(w)\Delta t$. Combining the growth into and out of the interval with the mortality gives the total change in the number of individuals within the size range $\Delta N\Delta w$ as

(continued)

$$\Delta N \Delta w = g\left(w - \frac{\Delta w}{2}\right)N\left(w - \frac{\Delta w}{2}\right)\Delta t - g\left(w + \frac{\Delta w}{2}\right)N\left(w + \frac{\Delta w}{2}\right)\Delta t$$

$$- N(w)\mu(w)\Delta w \Delta t. \tag{4.2}$$

(Box 4.2 *continued*)

As the size range is short, the growth at $w \pm \Delta w/2$ can be approximated by a linear expansion: $g(w \pm \Delta w/2) \approx g(w) \pm g'(w)\Delta w/2$, where $g'(w)$ is the derivative with respect to w. Similarly, for the spectrum: $N(w \pm \Delta w/2) \approx N(w) \pm N'(w)\Delta w/2$. Inserting these expansions into eq. 4.2 and simplifying gives (most of the terms cancel out)

$$\frac{\Delta N}{\Delta t} = -g'(w)N(w) - N'(w)g(w) - N(w)\mu(w). \tag{4.3}$$

Taking the limit $\Delta t \to 0$, and recognizing that $g'(w)N(w) + N'(w)g(w) = \partial(g(w)N(w))/\partial w$, gives the McKendric–von Foerster equation (eq. 4.1).

(numbers per weight), which is the size-based equivalent of an age distribution in classic age-based demography. The population size spectrum has the same characteristics as the community size spectrum in chapter 2 (box 2.1), except that it represents only individuals in a single population. In steady state, which is considered in this chapter, the time derivative in the first term of eq. 4.1 disappears and the partial differential equation reduces to an ordinary differential equation

$$\frac{dg(w)N(w)}{dw} = -\mu(w)N(w). \tag{4.4}$$

Knowing growth $g(w)$ and mortality $\mu(w)$, eq. 4.4 can be solved for the population size spectrum $N(w)$.

The growth rate of an individual was described in chapter 3 in two forms: the von Bertalanffy growth model (eq. 3.12)

$$g_{vb}(w) = Aw^n\left[1 - \left(\frac{w}{W_\infty}\right)^{1-n}\right], \tag{4.5}$$

where W_∞ is the asymptotic size and A the growth coefficient, and as the more complicated biphasic growth equation (eq. 3.18)

$$g_{bp}(w) = Aw^n\left[1 - \psi_m\left(\frac{w}{\eta_m W_\infty}\right)\left(\frac{w}{W_\infty}\right)^{1-n}\right], \tag{4.6}$$

where the function ψ_m describes maturation at size $\eta_m W_\infty$ (eq. 3.15). Using the von Bertalanffy growth model allows us to arrive at analytical solutions of the size spectrum that would become too complex with the biphasic growth model. In the interest of keeping things simple, I will also explore an even simpler growth model: $g_j(w) = Aw^n$. This model corresponds to juvenile growth in the biphasic growth equation when $\psi_m(w/w_m) = 0$. The analytical solutions are less accurate than the numerical solutions of the biphasic growth model, but they provide a clearer view of how population-level quantities, such as spawning stock biomass and reproductive output, vary systematically between species as a function of asymptotic size.

The mortality needed to solve the McKendrick–von Foerster equation equation (eq. 4.4) is the predation mortality inflicted by larger individuals in the community. In chapter 2, we saw how predation mortality followed metabolic scaling $\mu_p(w) = \Phi_p f_0 h w^{n-1}$ (eq. 2.22), where Φ_p was a complicated constant, f_0 the expected feeding level, and h the coefficient for maximum consumption rate. This relation shows that predation mortality is proportional to w^{n-1} and to the consumption rate $f_0 h$. Instead of bringing all these constants along, I write the mortality in a simpler form

$$\mu_p(w) = aAw^{n-1}, \tag{4.7}$$

where A is the same growth coefficient used in eqs. 4.5 and 4.6, and the new dimensionless parameter a is the *physiological mortality*. With $n = 3/4$, the mortality is declining with body size with exponent $-1/4$ and proportional to the growth coefficient A. The physiological mortality a turns out to be the most important parameter for describing fish demography and reproduction for reasons that will soon become abundantly clear, as will the reasons for writing the mortality as proportional to the growth coefficient. I will return to this parameter at length in section 4.4, but for now the focus is finding the population size spectrum.

Solving the McKendric–von Foerster equation for $N(w)$ is fairly straightforward when the simple growth equation $g(w) = Aw^n$ is used. As growth and mortality are both power-law functions, it is reasonable to make the *ansatz* that the spectrum is also a power-law solution: $N(w) = \mathscr{C}w^l$. $\mathscr{C}$ and l are constants representing the total abundance and the scaling of the spectrum that we aim to determine from the von Foerster equation. Inserting the power-law solution together with the growth and mortality (eq. 4.7) into the von Foerster equation (eq. 4.4) gives

$$\frac{\mathrm{d}Aw^n \mathscr{C}w^l}{\mathrm{d}w} = -aAw^{n-1}\mathscr{C}w^l \Leftrightarrow \frac{\mathrm{d}w^{n+l}}{\mathrm{d}w} = -aw^{n-1+l}. \tag{4.8}$$

We immediately see that the growth coefficient A cancels out—this is one reason why it is clever to formulate the mortality as in eq. 4.7. Evaluating the derivative on the left-hand side gives

BOX 4.3

ANALYTICAL SOLUTIONS TO THE SIZE SPECTRUM

To solve the steady-state McKendric–von Foerster equation (eq. 4.4), we must transform it to a form where it can be integrated (Beyer, 1989; Andersen et al., 2015). This can be achieved by observing that

$$\frac{d\ln(g(w)N(w))}{dw} = \frac{1}{g(w)N(w)}\frac{dg(w)N(w)}{dw}. \tag{4.9}$$

Isolating $dg(w)N(w)/dw$ and inserting in eq. 4.4 gives

$$\frac{d\ln(g(w)N(w))}{dw} = -\frac{\mu(w)}{g(w)}. \tag{4.10}$$

This equation can readily be solved by integrating on both sides

$$\int_{w_R}^{w} d\ln(g(w)N(w)) = -\int_{w_R}^{w}\frac{\mu(\omega)}{g(\omega)}d\omega. \tag{4.11}$$

Evaluating the integral on the right-hand side and taking the exponential on both sides gives

$$\frac{g(w)N(w)}{g(w_R)N(w_R)} = \exp\left[-\int_{w_R}^{w}\frac{\mu(\omega)}{g(\omega)}d\omega\right] = P_{w_R \to w}, \tag{4.12}$$

where I have identified the exponential term on the right-hand side as the survival from w_R to w, $P_{w_R \to w}$. The term in the denominator on the left-hand side, $g(w_R)N(w_R)$, is the flux of recruits at size w_R—this is the recruitment, R. Rearranging gives

$$\frac{N(w)}{R} = \frac{1}{g(w)}P_{w_R \to w}. \tag{4.13}$$

The integral determining the survival must in general be solved numerically (box 4.4), but two special cases allow analytical solutions:

Case I: Juvenile Growth, $g(w) = g_j(w) = Aw^n$ Inserting the mortality $\mu = aAw^{n-1}$, the integral in the survival can be directly integrated to give

$$P_{w_1 \to w_2} = \exp\left[-\int_{w_1}^{w_2}\frac{aA\omega^{n-1}}{A\omega^n}d\omega\right] = \left(\frac{w_2}{w_1}\right)^{-a}, \tag{4.14}$$

and the spectrum becomes

$$\frac{N(w)}{R} = \frac{w_R^a}{A}w^{-n-a} \quad \text{for} \quad g(w) = Aw^n. \tag{4.15}$$

(continued)

(Box 4.3 *continued*)

We can calculate the spawning stock biomass B_{SSB} (the biomass of adults) as the integral over the number spectrum from size at maturation $\eta_m W_\infty$ to the asymptotic size W_∞

$$\frac{B_{SSB}}{R} = \int_{\eta_m W_\infty}^{W_\infty} \frac{N(w)w}{R}\, dw = \frac{w_R^a}{A} \frac{1 - \eta_m^{2-n-a}}{(2-n-a)} W_\infty^{2-n-a}. \tag{4.16}$$

Case II: Von Bertalanffy Growth, $g(w) = g_{vb}(w)$ In this case, the integral in the survival evaluates to

$$P_{w_1 \to w_2} = \left(\frac{w_2}{w_1}\right)^{-a} \left[\frac{1 - (w_2/W_\infty)^{1-n}}{1 - (w_1/W_\infty)^{1-n}}\right]^{\frac{a}{1-n}}, \tag{4.17}$$

that is, the same solution as earlier (eq. 4.14), with a correction given in the square brackets. The expression can be simplified by observing that if $W_\infty \gg w_1$, the denominator in the square brackets is ≈ 1. Inserting the von Bertalanffy growth formula from eq. 4.5, the spectrum can be approximated as

$$\frac{N(w)}{R} \approx \frac{w_R^a}{A} w^{-n-a} \left[1 - \left(\frac{w}{W_\infty}\right)^{1-n}\right]^{\frac{a}{1-n}-1} \quad \text{for} \quad W_\infty \gg w_R. \tag{4.18}$$

In this case, the spawning stock biomass B_{SSB} cannot be solved in general, but for $n = 3/4$, it is

$$\frac{B_{SSB}}{R} = \frac{w_R^a}{6A} \left(-24B[\eta_m^{1/4}; 5 - 4a, 4a] + \Gamma[5-4a]\Gamma[4a]\right) W_\infty^{5/4-a}, \tag{4.19}$$

where Γ is the gamma function, and $B[\cdot]$ is the incomplete beta function.

$$(n+l)w^{n+l-1} = -aw^{n+l-1}. \tag{4.20}$$

Solving for the exponent gives $l = -n - a$, so the spectrum is

$$N(w) = \mathscr{C}w^{-n-a}, \tag{4.21}$$

with $\mathscr{C}$ yet undetermined. Interestingly, we see that the physiological mortality a enters together with the metabolic exponent n in the exponent.

Boundary Condition

So far, we have solved only for the scaling of the size spectrum and left the constant $\mathscr{C}$ unspecified. This constant determines the total abundance and biomass of the population. The constant can be found by considering the flux of individuals R

at the smallest body size w_R. These individuals are called *recruits*. The flux of recruits (numbers per time) is the same as the flux of individuals growing into the first size group in fig. 4.1. We can therefore use the recruitment flux to determine the growth flux $g(w_R)N(w_R)$ at the boundary

$$g(w_R)N(w_R) = R. \qquad (4.22)$$

This is the boundary condition to the McKendrik–von Foerster equation. Both growth rate and spectrum at the boundary are known: the growth rate is $g(w_R)$ and the spectrum at the boundary $N(w_R)$ is $\mathscr{C} w_R^{-n-a}$. Inserting those two expressions

---

BOX 4.4

NUMERICAL SOLUTION OF THE SURVIVAL

Solving the size spectrum with arbitrary growth and mortality functions requires a numerical solution of the survival $P_{w_1 \to w_2}$. This solution is achieved in three steps, as follows (Andersen et al., 2015):

1. First construct a series of m weight classes w_i, logarithmically distributed between w_R and W_∞,

$$w_i = \exp\left[\ln(w_R) + (i-1)\Delta\right], \qquad (4.23)$$

   where

$$\Delta = \frac{\ln W_\infty - \ln w_R}{m-1}. \qquad (4.24)$$

   For the calculations presented in this chapter, $m = 1000$ was used though less will in most cases be sufficient.

2. Define the physiological mortality at each grid point, a_i, as mortality divided by specific growth

$$a_i = \frac{\mu(w_i)}{g(w_i)} w_i. \qquad (4.25)$$

   For the calculations in this chapter, $a_i = a$ is just a constant, but in later chapters mortality also includes fishing mortality.

3. Approximate the survival as

$$P_{w_R \to w} = P_{1 \to i} \approx \exp\left[-\Delta \sum_{j-2}^{i} a_{j-1}\right] \quad \text{for} \quad i \geq 2, \qquad (4.26)$$

   with $P_1 = 1$. In most higher level programming languages, the sum can be implemented as a cumulative sum, for example, using the command "cumsum" in R or Matlab.

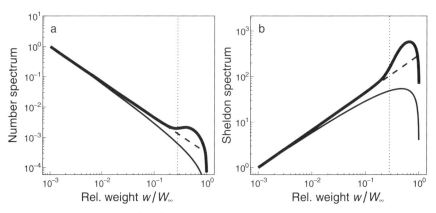

FIGURE 4.2. Size spectra as a function of individual size relative to asymptotic size; (a) number spectrum $N(w)$; (b) Sheldon spectrum $N(w)w^2$. The spectra are scaled to have value one at the left side. The thick line shows the spectrum using the biphasic growth equation (eq. 4.6), the thin line shows the analytical solution to the spectrum using the von Bertalanffy growth equation (eq. 4.27), and the dashed line shows the scaling of the juvenile spectrum, w^{-n-a} and w^{2-n-a}, respectively. The vertical dotted lines are at the size of maturation.

in the boundary condition gives $\mathscr{C} = Rw_R^{a+n}/g(w_R)$, which can be reduced to $\mathscr{C} = Rw_R^a/A$ for the simple growth equation $g(w_R) = Aw_R^n$.[1]

Solving the McKendrick–von Foerster equation using the von Bertalanffy growth model (eq. 4.5), including the boundary condition, gives (box 4.3)

$$N(w) = R\frac{w_R^a}{A}w^{-n-a}\left[1 - \left(\frac{w}{W_\infty}\right)^{1-n}\right]^{\frac{a}{1-n}-1} \quad \text{for} \quad W_\infty \gg w_R. \quad (4.27)$$

The first term in eq. 4.27 scales as w^{-n-a}, just as the simple solution. The term in the square brackets is close to 1 for juveniles, when $w \ll W_\infty$. As individuals mature, the term in the square brackets becomes increasingly important, and the solution diverges toward zero (or toward infinity if $a < 1 - n$).

The three solutions—the simple solution, the one based on von Bertalanffy growth in eq. 4.27, and the numerical solution using the biphasic growth equation—are plotted in fig. 4.2. The solution with biphasic growth differs from the two other solutions by an increase in abundance around the size of maturation. As available energy is being directed toward reproduction, there is less available for growth, and growth declines. This leads to a pile-up of individuals in the

---

[1] Strictly speaking, this approximation is valid only for the solutions with the simple growth model and the biphasic growth model. However, considering that $w_R \ll W_\infty$, it is also a very good approximation for the solution with the von Bertalanffy growth model (eq. 4.27).

size range where growth slows down, much as the density of cars on a highway increases when drivers brake before road work.

What Is the Cohort Biomass?

So far, so good. Now, let's use the solutions for something. To keep things simple, I will mainly base the derivations on the simple solution based on juvenile growth, $N(w) \propto w^{-n-a}$.

First, I will derive the biomass of a cohort, following Law et al. (2016). In classic age-based life tables, the biomass of a cohort is readily calculated as the abundance in an age class multiplied by the weight-at-age. In the size-based approach, again, growth has to be factored in. The biomass of a cohort per egg is the survival multiplied by the weight of individuals

$$B_{cohort} = P_{w_R \to w} w. \tag{4.28}$$

Survival is found as the solution to

$$\frac{dP}{dt} = -\mu P, \tag{4.29}$$

which is:

$$P(t) = P(0) \exp\left[-\int_0^t \mu(t)\, dt\right] = P_{0 \to t}. \tag{4.30}$$

Changing the integration variable from time to weight using $dt = (dt/dw)dw = (1/g(w))dw$ and new limits w_R (at time $t = 0$) and w at time t gives

$$P_{w_R \to w} = \exp\left[-\int_{w_R}^w \frac{\mu(w)}{g(w)}\, dw\right] = \left(\frac{w}{w_R}\right)^{-a}, \tag{4.31}$$

where the last step is found by inserting the mortality $\mu = aAw^{n-1}$ (eq. 4.7) and juvenile growth $g(w) = Aw^n$. The biomass of a cohort then becomes

$$B_{cohort} = w_R^a w^{1-a}. \tag{4.32}$$

As we will see later, the physiological mortality is less than 1, so the biomass of a cohort increases with size (fig. 4.3). When individuals reach maturity their growth rate slows down and the biomass of the cohort begins to decrease. The increase in cohort biomass with size and age until maturation is a classic result in fisheries science. It is the basis of mesh size regulations in fisheries because it predicts that in order to maximize yield one should wait for the cohort to reach the peak of biomass before harvesting commences—more on that in chapter 5.

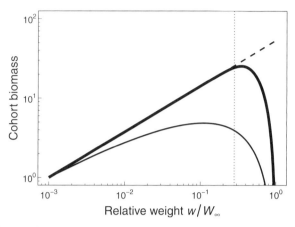

FIGURE 4.3. Cohort biomass relative to initial cohort biomass as a function of body size. The thick line shows cohort biomass using the biphasic growth equation (eq. 4.6), the thin line shows the analytical solution using the von Bertalanffy growth equation (eq. 4.5), and the dashed line shows the approximation using the simple growth model (eq. 4.32). The vertical dotted line is at the size of maturation.

A commonly used measure of population size is the spawning stock biomass, B_{SSB}. The spawning stock biomass represents the total biomass of the adult population. The spawning stock biomass is the integral of the biomass spectrum from the size at maturation $\eta_m W_\infty$ to asymptotic size W_∞

$$B_{SSB} = \int_{\eta_m W_\infty}^{W_\infty} N(w)w\,dw = R\frac{w_R^a}{A}\frac{(1-\eta_m)^{2-n-a}}{2-n-a}W_\infty^{2-n-a} \quad \text{for} \quad g(w) = Aw^n.$$
(4.33)

Using the other growth models will change the terms in front of the term with W_∞, but the scaling with asymptotic size will be the same (eqs. 4.16 and 4.19). The scaling of B_{SSB} with asymptotic size is $B_{SSB} \propto W_\infty^{2-n-a} \approx W_\infty^{0.83}$ (for $a = 0.42$). The spawning stock biomass increases rapidly with asymptotic size.

Spawning stock biomass per recruit B_{SSB}/R is an increasing function of asymptotic size because the exponent $2 - n - a > 0$. If a small species and a large species spawn the same number of eggs, then the spawning stock biomass of the large species will eventually be larger than the small species. For example, if the small species is a forage fish with asymptotic size $W_\infty = 10$ g and the large species a demersal species with asymptotic size $W_\infty = 10$ kg, the spawning stock biomass of the large species will be $1000^{2-n-a} \approx 300$ times larger, if $a = 0.42$. But in reality, we see that forage fish have huge stock sizes, certainly larger than stocks of large-bodied species. How is that reconciled with their lower B_{SSB}/R? Part of the answer to that conundrum lies in differences in the strength of density-dependent regulation between stocks of large and small bodied species, which I'll turn to next.

4.2   REPRODUCTION, RECRUITMENT, AND DENSITY DEPENDENCE

Recruitment is the technical term in fisheries science for the individuals that survive the early life stages to become available to the fishery. The age (or size) at recruitment is often specified in vague terms (if at all), but it is usually at older and larger stages the larger the asymptotic size. What matters though, is that all density-dependent effects are assumed to occur before recruitment—that is, at sizes smaller than w_R. In this section, I will introduce the standard method to deal with prerecruit density dependence and discuss the implications and the assumptions behind this practice at length in the following section.

First, though, I will disregard density-dependent effects to determine the reproductive output R_p from the population. The recruitment flux is found by combining the individual reproductive output from chapter 3 with the abundance of individuals from the size spectrum.

The population's reproductive output is produced by the adults. In chapter 3, we found that each adult had a reproductive output of $R_{egg} = \varepsilon_{egg} A W_\infty^{n-1} w$ (biomass per time; eq. 3.19). Integrating the individuals' reproductive output over the adult population gives the populations reproductive output as numbers of hatched larvae per time

$$R_p = \frac{\varepsilon_R}{w_0} P_{w_0 \to w_R} \int_{w_R}^{W_\infty} \psi_m \left(\frac{w}{w_m} \right) R_{egg} N(w) w \, dw \qquad (4.34)$$

$$= \frac{\varepsilon_R \varepsilon_{egg}}{w_0} P_{w_0 \to w_R} A W_\infty^{n-1} B_{SSB}. \qquad (4.35)$$

The division by the egg weight w_0 is used to change the units from biomass to numbers. The reproductive output is discounted by a *recruitment efficiency* ε_R that encompasses egg survival and is further discounted by survival from egg size w_0 to the size of recruitment $P_{w_0 \to w_R} \approx (w_R/w_0)^{-a}$ (eq. 4.31).

We can use the simple solution of spawning stock biomass (eq. 4.33) $B_{SSB} \propto R w_R^a W_\infty^{2-a-n}$ to determine how the reproductive output scales with asymptotic size: $R_p \propto \varepsilon_R \varepsilon_{egg} R W_\infty^{1-a}$. Again, we see an increasing function of W_∞ as $a < 1$: larger species make more eggs per recruit than smaller species. I will return to this property at the end of this section.

I have so far dodged the need to consider density dependence. If we use the reproductive output as the recruitment, the population will grow exponentially without bounds (assuming $R_p/R > 1$). However, a population at equilibrium neither grows nor decays in abundance. The equilibrium is obtained by density-dependent effects that regulate the number of recruits per recruit to 1. That an effect is "density dependent" means that it varies as a function of the abundance, or the density, of the population. Examples of density-dependent processes are a growth rate that decreases as the abundance increases owing to competition for food, a mortality that increases with abundance owing to starvation or cannibalism, or a reduction in individual reproductive output owing to high abundance of adults or eggs. All these effects reduce the recruits per recruit until $R_p/R = 1$ and the population is in equilibrium. Such population regulating effects need to be accounted for.

The simplest representation of density-dependent regulation is a *stock-recruitment* relation that describes recruitment R as a function of the reproductive output R_p. The most common stock-recruitment relation is the Beverton-Holt function

$$R = R_{\max} \frac{R_p}{R_p + R_{\max}}. \tag{4.36}$$

This function increases linearly with the reproductive output R_p and saturates at the maximum recruitment $R_{\max}$. $R_{\max}$ acts as the carrying capacity of the population. $R_{\max}$ is set by environmental factors defined outside the theory developed here, such as the size of the suitable habitat for the species or the amount of food in the early life history in the size range w_0 to w_R. I'll return to $R_{\max}$ in chapter 11.

Getting an intuitive understanding of where a given stock is positioned on the *x*-axis of the stock-recruitment relationship at equilibrium is tricky. Everything appears connected, which makes it difficult to disentangle cause from effect: the recruitment enters into the equation of the spawning stock biomass B_{SSB} (eq. 4.16), which in turn determines reproductive output (eq. 4.35) and then recruitment (eq. 4.36). It is easy to accept that the asymptotic-size scaling of spawning stock biomass and the egg production per recruit can be calculated from the size spectrum of the population. However, to get the total spawning stock biomass and the total egg production we must multiply by the number of recruits, which depends on the recruitment itself. Mathematically, we know only the B_{SSB} and egg production relative to recruitment—for example, B_{SSB}/R from eqs. 4.16 and 4.19. The egg production can be written as $R_p/R = \alpha B_{SSB}/R$, where the variable α contains all the constants in eq. 4.35: $\alpha = \varepsilon_R \varepsilon_{egg} P_{w_0 \to w_R} A W_\infty^{n-1}/w_0$. Inserting in the stock-recruitment relation eq. 4.36 and rearranging gives

$$\frac{R}{R_{\max}} = 1 - \frac{1}{\alpha(B_{SSB}/R)}, \tag{4.37}$$

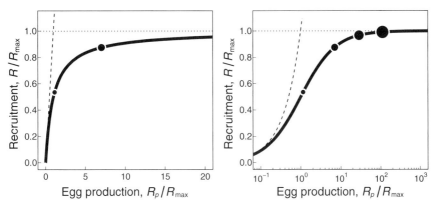

FIGURE 4.4. The Beverton-Holt stock recruitment function as a function of the reproductive output relative to maximum recruitment, plotted with linear and logarithmic scales on the x-axis. The dashed line represents the slope at the origin, $R = R_p$. The black dots represent species with increasing asymptotic size (and increasing point size): 10 g, 100 g, 1 kg, and 10 kg.

and

$$\frac{R_p}{R_{max}} = \alpha \frac{B_{SSB}}{R} - 1. \tag{4.38}$$

If $\alpha B_{SSB}/R < 1$, the recruitment becomes negative. This may seem odd, but it simply shows that the population collapses when one recruit produces less than one recruit itself. If $\alpha(B_{SSB}/R) > 1$, the recruitment is positive but it never becomes larger than the maximum recruitment.

Fig. 4.4 makes it clear that larger bodied species have a higher R_p/R_{max} than smaller species: larger species are further to the right on the stock-recruitment curve, whereas smaller species are closer to the origin. In other words, large-bodied species are expected have an almost flat stock-recruitment curve, because the rising part of the curve will rarely be observed. Therefore, larger species are often managed without consideration of a stock-recruitment relationship, under the assumption that recruitment is just constant. Small species, on the other hand, are close to the linear rising part of the stock-recruitment curve. Such species are often said to not have a stock-recruitment relationship, and fisheries advice is provided under the assumption that recruitment is proportional to egg production.

A traditional measure of the reproductive capacity and health of a fish stock is the "eggs-per-recruit," $R_0 = R_p/R$. This measure is equivalent to the lifetime reproductive output R_0 in life-history optimization theory, where it is used as a proxy of fitness. R_0 is a nondimensional measure that states how many eggs the average hatched larva produces during its entire life. Most larvae, of course, die, but that is compensated by the enormous number of offspring produced by the few lucky survivors to adulthood. R_0 weighs the huge mortality of juveniles against the

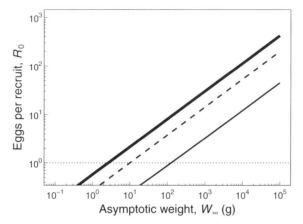

FIGURE 4.5. Eggs per recruit R_0 as a function of asymptotic size. The thick line shows the full numerical solution calculated with the biphasic growth equation (eq. 3.18), the thin line shows the analytical solution using the von Bertalanffy growth equation (eq. 4.5), and the dashed line shows the approximation using juvenile growth, $R_0 \propto W_\infty^{1-a}$. The horizontal dotted line at R_0 indicates the lower limit for persistence of the population at $R_0 = 1$.

enormous reproductive capacity of adults. If $R_0 = 1$, then the population just balances; every recruit produces one viable egg on average. If $R_0 > 1$, the population increases and vice versa. The value of R_0 therefore indicates how close the population is to the extinction threshold at $R_0 = 1$. R_0 is simply the ratio between the reproductive output R_p and the recruitment flux R (Andersen et al., 2008)

$$R_0 = \frac{R_p}{R} = \frac{\varepsilon_R \varepsilon_{\text{egg}}}{w_0} P_{w_0 \to w_R} A W_\infty^{n-1} \frac{B_{\text{SSB}}}{R} \propto \varepsilon_R W_\infty^{1-a}. \qquad (4.39)$$

The first part of the relation is found using eq. 4.35, and the scaling of B_{SSB} from eq. 4.33 is used for the last bit. This result is again based on the approximation of the spectrum calculated from the juvenile growth, but the result is the same for the von Bertalanffy equation and the biphasic growth equation up to a constant of proportionality (Andersen et al., 2008) and an almost imperceptible correction for small values of W_∞ (fig. 4.5).

The calculation of R_0 is useful because it quantifies the amount of density-dependent regulation needed to achieve equilibrium. Because R_0 is an increasing function of asymptotic size, larger species need a higher degree of density-dependent regulation (a larger reduction in R_0) than smaller species. Such differences in the strength of density dependence between small and large species was observed in the Barents Sea. There, the large-species cod and haddock have strong density-dependent regulation, whereas the small capelin has little regulation (Dingsør et al., 2007). There are also indications that this pattern exists broadly: Goodwin et al. (2006) made a meta-analysis of stock-recruitment relationships and found that large-bodied and late-maturing species had strong density-dependent recruitment. In contrast, small-bodied, early-maturing fish had high annual recruitment variability and weak density-dependent control.

Eq. 4.39 gives two important insights concerning the role of the asymptotic size on population fitness: it is bad to be small, and it is good to be big. First, at some small asymptotic size the lifetime reproductive output will be less than one. This means that one hatched larva on average will produce less than one larva itself. Consequently, the population is unable to support itself and will slowly but surely collapse. For the parameters chosen here, this happens at an asymptotic size around 1 g. Indeed, many small fish species employ special measures to boost offspring survival by increasing the recruitment efficiency ε_R: they are mouth brooders (for example, cichlids), they guard their eggs (for example, gobies or sticklebacks), or they bear live offspring instead of eggs. In contrast, most larger species just spawn their eggs freely in the water column or on the seabed and leave them to fend for themselves. Second, since larger species have a higher R_0 than small species, larger species are increasingly "fit" in the sense that they produce many eggs per recruit. Deriving this result was a surprise to me. I had expected that the number of eggs per recruit would decrease for larger species because their offspring are increasingly unlikely to survive to adulthood. However, the low survival is more than compensated by the larger fecundity of the large fish. This surprising result is the cornerstone of the remarkable resilience of fisk stocks to fishing that we will see in the next chapter.

4.3   WHY USE A STOCK-RECRUITMENT RELATION?

I have used a stock-recruitment relationship to represent the process of density dependence in a population. This choice follows the tradition in fisheries science where density dependence is almost exclusively represented by stock-recruitment relations. I have developed the results using one particular stock-recruitment relationship but an entire zoo of different relationships has been described (Shepherd, 1982). They all share some qualitative features: recruitment increases as a function of the spawning stock biomass, and they either reach an asymptotic maximum recruitment or have decreasing recruitment at high spawning stock biomasses.

Stock recruitment relations have been the subject of lavish attention: Which stock-recruitment relationship should be used in a particular case? What is the value of the central parameter α that relates the initial slope to the spawning stock biomass? Is there depensation—that is, lowered recruitment at small stock sizes, also known as an Allee effect (Myers et al., 1995)? Or is there overcompensation—that is, lowered recruitment at high stock sizes (Myers, 2001)? Empirical data on recruitment are plagued by large uncertainties: the data supporting the fitted stock-recruitment relationships are typically of exceptionally poor quality. One question, however, is rarely asked: Why do we use stock-recruitment relationships to represent density dependence? Or, in other words, does density dependence really occur only early in life, as implied by the stock-recruitment relation? Why does it not occur later in life among adults? A pragmatic answer to the first question is simple: we use stock-recruitment relationships because they are convenient and easy. The stock-recruitment relationship parameterizes all of the complexities of density-dependent regulation into one simple function. However, the validity of that approach hinges on whether density dependence really happens early in life. I will address this question in detail in chapter 10, but until then I will assume that density dependence is well described by a stock-recruitment relationship.

4.4 WHAT IS THE PHYSIOLOGICAL MORTALITY?

The development of the demographic model saw the birth of a new parameter that appears in almost every important relation involving demography: the physiological mortality constant a. The concept was first introduced by Beyer (1989) in his work on size-based theory and recruitment. He defined the physiological mortality as the ratio between mortality and weight-specific available energy for growth or reproduction E_a/w:

$$a = \frac{\mu(w)}{E_a(w)/w}. \tag{4.40}$$

With mortality $\mu \propto w^{n-1}$ (dimensions per time) and the available energy given as $E_a(w) = Aw^n$ (mass per time), we see that a is indeed dimensionless and independent of body size. The physiological mortality is closely related to the ratio μ/g that appears centrally in the calculation of survival (eq. 4.31) and the spectrum. For juveniles, all available energy is used for growth, so $a = (\mu/g)/w$. The ratio μ/E_a is also central in behavioral ecology where Gilliam's rule (Gilliam and Fraser, 1987) states that an individual will always strive to minimize μ/E_a (at least in a stable environment; see Sainmont et al. 2015). Minimizing μ/E_a is the same as minimizing a. As a always appears as a negative term (in the survival $\propto w^{-a}$, the spectrum $\propto w^{-n-a}$, and $R_0 \propto W_\infty^{1-a}$), lowering a generally leads to a better population-level performance.

In the following, I will develop two estimates of a, one based on the size spectrum theory from chapter 2 and the energy budget in section 3.3, and one based on observations of growth and mortality among fish species.

The size spectrum theory in chapter 2 gave us a mechanistic description of mortality from the predator-prey interactions in the marine size spectrum (eq. 2.22): $\mu(w) = \Phi_p f_0 h w^{n-1}$, where $\Phi_p \approx 0.17$ was a constant describing predator-prey size preferences, $f_0 \approx 0.6$ the expected feeding level, and h the coefficient of maximum consumption rate (table 2.2). Inserting that description of mortality in the definition of a in eq. 4.40 gives $a = \Phi_p f_0 h / A$. In chapter 3, the growth coefficient A was related to assimilation efficiency $\varepsilon_a \approx 0.6$ and standard metabolism represented by the critical feeding level $f_c \approx 0.2$ as $A = \varepsilon_a h(f_0 - f_c)$ (eq. 3.31). Inserting the relations for μ and A in the definition of the physiological mortality eq. 4.40 gives

$$a = \frac{\Phi_p f_0}{\varepsilon_a(f_0 - f_c)} \approx 0.424. \tag{4.41}$$

This constitutes a mechanistic derivation of the physiological mortality on the basis of parameters related to individuals' prey size preferences, as represented by Φ_p, and physiology as represented by ε_a, f_0, and f_c. The estimated values of the parameters are fairly uncertain—in particular, f_0 and f_c—which is why this estimate of a should be considered only a rough approximation.

An empirical estimate can be made by realising that the physiological mortality is closely related to the M/K life-history parameter used in fisheries science and life-history theory (Beverton and Holt, 1959; Charnov et al., 2001). M/K is the ratio between the adult natural mortality M and the von Bertalanffy growth constant K (section 3.1). Both quantities have dimensions time^{-1} and their ratio M/K is dimensionless. In their work on the demography of fish stocks, Beverton and Holt were clearly aware of the importance of the ratio of these two parameters to describe demography, and Beverton kept trying to uncover systematic patterns in M and K. He realized that their ratio was fairly constant, varying roughly between 0.5 and 2 with a mean around 1 (Beverton, 1992). The connection between M/K and the physiological mortality a can be uncovered by defining adult mortality M as the size-based mortality (eq. 4.7) at the size at maturation $\eta_m W_\infty$: $M = \mu_p(\eta_m W_\infty) = a A \eta_m^{n-1} W_\infty^{n-1}$. In the previous chapter, we derived the relation between the von Bertalanffy growth constant K and asymptotic weight as (box 3.2): $K = A c^{-1/3} w^{n-2/3} L_\infty^{-1}/3$. Defining the value of K at the size at maturation $w = \eta_m W_\infty$, the ratio M/K becomes (Andersen et al., 2009a)

$$\frac{M}{K} = 3a\eta_m^{-1/3} \Leftrightarrow a = \frac{1}{3}\frac{M}{K}\eta_m^{1/3} \approx 0.22\frac{M}{K}. \tag{4.42}$$

The physiological mortality is proportional to the M/K life-history parameter with a coefficient of proportionality around 0.2. We can now estimate a from observations of M and K. Reliable estimates of natural mortality M are, however, notoriously difficult to come by for free-living fish populations (Gislason et al.,

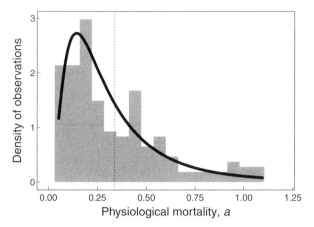

FIGURE 4.6. Histogram of values of the physiological mortality a. The values are based on the compilation of M and K by Gislason et al. (2010) (only for adults and with values of $a > 1.25$ omitted). The simple approximation $a \approx 0.22(M/K)$ is used to calculate a (eq. 4.42). The vertical dashed line shows the mean value ($a = 0.34 \pm 0.25$). The thick line is a fit to a log-normal distribution with a mean of 0.26.

2010). One nagging issue is that most studied populations are also fished, and the natural mortality cannot easily be disentangled from the fishing mortality. The compilation of mortality data in fig. 4.6 shows a large variation in physiological mortalities.[2] The mean is $a \approx 0.34$, fairly close to the theoretical prediction of the average value at 0.425 (eq. 4.41).

It is now clear that the physiological mortality a and the ratio between adult mortality and the von Bertalanffy growth constant M/K are two sides of the same coin. The only difference is that the physiological mortality covers all life stages and not just adults.

Taking a step back, the physiological mortality (and M/K) can be perceived as representing two aspects of life in the ocean: a link between growth and mortality, and a neutral trade-off between fast-slow life-history strategies. The link between growth and mortality was already evident in chapter 2, where we considered the predator-prey relations in the entire size spectrum: every inch of growth in larger

---

[2] The data by Gislason et al. (2010) used in the figure include some data points from juvenile individuals, which I have ignored. One important result of their analysis was that juvenile mortality is higher than expected by metabolic scaling arguments—that is, that mortality scales with an exponent steeper (more negative) than $n - 1$, as also shown by McGurk (1986). This increased mortality among juveniles is commensurate with the observation of increased density-dependent mortality among juveniles (Myers and Cadigan, 1993). The observation of a steeper scaling of mortality than $n - 1$ among juveniles conflicts with the prediction from chapter 2. The conflict is, however, apparent only if the increased mortality is interpreted as density-dependent mortality. The total mortality should be composed of a density-independent mortality—for example, eq. 4.7—and a density-dependent mortality that is captured by the stock-recruitment relation.

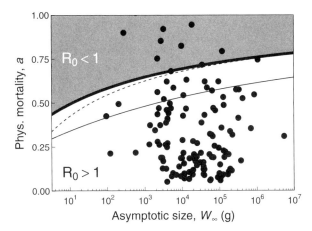

FIGURE 4.7. The value of the physiological mortality a where the population can just sustain itself as a function of asymptotic size ($R_0 = 1$). Numerical solution (thick line), the analytical approximation (thin line), and approximated from the juvenile spectrum (dashed line). The points are from the same data set as in fig. 4.6. Note several points outside the range where populations are predicted to persist—this reflects partly that the theoretical prediction is based only on average values of the constants entering it, and partly the difficulty in accurately estimating mortality.

organisms is fueled by a corresponding mortality on smaller organisms. That link leads to the relation between growth and mortality reflected in the theoretical prediction of the physiological mortality in eq. 4.41. While the link is evident on average, specific populations might have different values of a than implied by the balance between growth and mortality (evident in the large spread of values in fig. 4.6).

A population may, for example, have established itself in a particularly favorable habitat with a lower value of a. Conversely, another population of the same species may live under circumstances that expose individuals to an above average predation risk or to a lower than average scope for growth, both leading to a higher value of a. A higher value of a makes the population more sensitive to external perturbations, environmental as well as anthropogenic, but as long as $R_0 > 1$ the population can persist. The physiological mortality of specific populations can therefore take on a range of values up to the level where the population will go extinct—that is, where $R_0 = 1$. The physiological mortality is therefore not a life-history parameter determined by physiology such as the growth constant A, but it is determined by the environment, and it may take a range of values between different populations of the same species. We have also seen that species with larger asymptotic size have a higher value of eggs per recruit R_0 (fig. 4.5). This means that they can tolerate a higher value of the physiological mortality before they go extinct. In practical applications, we can expect the physiological mortality to take a range of values, and that the variability of values between species is higher for species with a larger asymptotic size than species with a lower asymptotic size. Fig. 4.7 shows that the predicted difference between large and small species is fairly subtle and it cannot be seen in the very noisy data.

We can also perceive the relationship between growth and mortality represented in eq. 4.7 as a reflection of a trade-off in a life-history choice: a species with a given asymptotic size can adopt a live-fast-die-young strategy with high growth rates and short life span, or it can play it safe and with slow growth and a long life. The trade-off can even exist within different populations of the same species. In experiments on Atlantic silversides, Lankford et al. (2001) demonstrated how individuals with faster growth rates had substantially higher mortality rates than their slower growing cousins. With the trade-off implied by mortality being proportional to the growth coefficient A, survival to adulthood and life-time reproductive output of these different strategies are the same. Whether the species adopts the fast or the slow strategy has no impact on demography; only the ratio a matters for the size distribution (the juvenile number spectrum being $\propto w^{-n-a}$), survival to adulthood $\propto W_\infty^{-a}$, or eggs per recruit (lifetime reproductive output) $R_0 \propto W_\infty^{1-a}$. The trade-off implied by eq. 4.7 between fast and slow life-history strategies is therefore neutral; fast or slow species with the same asymptotic size will enjoy the same fitness. The growth rate does, however, matter for population growth rates—I'll treat this problem comprehensively in chapter 7.

4.5   SUMMARY

We now have a complete theory of population demography that can make general predictions of a fish population's demography and performance (equations and parameters are summarized in appendix A). We can use the theory for a specific stock by specifying all the parameters that enter into the results, or we can assume that all parameters, except the asymptotic size, take their average values. In this way, we can make general statements about how demography varies between fish population just by varying the asymptotic size.

This chapter involved a deep dive into the mathematics of population size spectra, with detours to the nature of density-dependent regulation and the ecological theory of the M/K parameter. The derivations of simple approximate solutions revealed important aspects of basic demography: the size structure of a population scales as $N(w) \propto w^{-n-a}$, the biomass of a cohort increases with size as w^{1-a}, the B_{SSB} per recruit scales with asymptotic size as W_∞^{2-n-a}, and eggs per recruit scale as $R_0 \propto W_\infty^{1-a}$. These scaling relationships were backed up by the complete analytical solutions based on the von Bertalanffy growth model and by the full numerical solutions of the biphasic growth model. The physiological basis of the growth model allowed us to go beyond basic demography and make predictions of reproductive output and recruitment. Here, the central result was that the degree of density-dependent regulation increases with asymptotic size.

The full description of a population's demography of course relies on other parameters than the asymptotic size. The most important are the physiological mortality a, the recruitment efficiency ε_R, and the maximum recruitment $R_{\max}$. The key parameter is clearly the physiological mortality a, which, together with the metabolic exponent n, plays a major role. In fact, a may be considered even more important than n, because the relations involving recruitment were determined only by a and not by n. Regarding recruitment, a central parameter is the recruitment efficiency ε_R, which characterizes the overall hatching success of eggs. This parameter represents environmental effects on recruitment outside the scope of the theory, and there is no simple mechanistic explanation for its value. The recruitment efficiency is expected to vary between stocks; some may have found a very suitable spawning habitat leading to a high recruitment efficiency or vice versa. There is no reason to expect a systematic variation in ε_R between species of different size—eggs and larvae of large species face the same challenges for survival as those from small species. An exception may be the small species who employ special strategies to increase ε_R, such as live offspring, and paternal or maternal care. I have not discussed the maximum recruitment $R_{\max}$ much. It characterizes the carrying capacity of the stock in the stock-recruitment relationship. As such, it represents the size of the habitat and is outside the scope of the theory.

Last, the growth coefficient A characterizes the slow-fast continuum of life histories. Somewhat surprisingly, perhaps, the theory showed that fitness, represented by the eggs per recruit R_0, does not depend on A. The demography is therefore neutral to changes in A: a fast-growing species will have the same size distribution and fitness as a slow-growing species. This prediction is contingent on the proportionality between growth and mortality in eq. 4.7. That is, an increase in growth rate leads to an increase in mortality. Growth and mortality might not always be connected in this manner. Take as an example the Baltic cod. In the 2000s, the condition of individuals in the stock began to decline (Eero et al., 2012). The decline is probably caused by a lower availability of sprat or benthic prey, leading to slower growth and possibly also increased susceptibility to parasites. The consequences would be declining growth (lower A) and increasing mortality. As the physiological mortality represents the ratio between growth and mortality, such a situation corresponds to increased physiological mortality, leading to a steeper spectrum and lower recruitment. Changes in the environment of a particular stock will therefore be reflected as changes in a.

CHAPTER FIVE

Fishing

How does fishing impact the demography and recruitment of a fish stock? Fishing targets all kinds of species, from small forage fish, such as anchovies or sand eel, to large-bodied demersal species, such as cod or saithe; from slow-growing redfish to fast-growing scombroids (tuna, swordfish, and so on). Clearly, the impact of fishing depends on the species—some species are very resilient to fishing, while others tolerate only little exploitation. How hard can a particular stock be fished before its productivity is compromised? And when is the stock's existence threatened?

Notice the difference in terminology between fisheries managers and scientists and ecologists: fisheries scientists refer to "stocks," while ecologists talk about "populations." Both terms refer to the same quantity—namely, a localized population of individuals from the same species. I will use *population* when I talk about fish in general, and use *stock* when I refer particularly to exploited populations.

The impact of fishing was addressed by Thomas Huxley in his inaugural address at the Fisheries Exhibition in London in 1882 (fig. 5.1). Having devoted some thought to the problem, Huxley famously concluded: "I believe . . . that all the great sea fisheries, are inexhaustible; that is to say, that nothing we do seriously affects the number of the fish." Huxley has often been quoted and ridiculed for the statement. He was, however, less categorical than the quote implies and also stated that, "I have no doubt whatever that some fisheries may be exhausted." Huxleys statement of inexhaustible fisheries was based on a fairly loose argumentation comprising anecdotal evidence, a superficial assessment of fishing mortality (less than 5 percent per year), and a consideration of natural mortality: "The great shoals are attended by hosts of dog-fish, pollack, cetaceans and birds, which prey upon them day and night, and cause a destruction infinitely greater than that which can be effected by the imperfect and intermittent operations of man." On this basis, he concluded: "any attempt to regulate these fisheries seems consequently, from the nature of the case, to be useless." Unfortunately, he was wrong. Already at the time of his address at the Fisheries Exhibition, the reality of overfishing had shown its face among the "great fisheries." The Atlantic halibut fishery had collapsed, and the U. S. Fish Commission was established 12 years earlier to figure out why fisheries in New England were declining.

FIGURE 5.1. Thomas Huxley addressing the audience at the opening of the Fisheries Exhibition in London 1882. *Source:* https:mathes.clarku.edu/huxley/SM5/fish.html.

If fisheries are to be regulated, what does a manager need to know? Fisheries management has inherited the utilitarian view espoused by Huxley. From that standpoint, the central question is: How much yield can be extracted from a given fish stock? And what is the optimal gear type to maximize yield? Realizing that a fish stock is not a factory where everything is under its owners' absolute control, but is influenced by an everchanging natural ecosystem, the manager also wants to know the consequences of changed environmental conditions: what happens if recruitment is compromised, if food conditions deteriorate, or if the losses to predation by larger fish or marine mammals change? These aspects are represented by fisheries reference points. A reference point condenses the information about the impact of fishing into a number that is used as either a target for management, such as a desired biomass, or an upper limit to the fishing mortality that should be avoided. Last, for the many stocks that are in a state of overfishing and rebuilding, the time scale of recovery is relevant; that is the focus of chapter 7.

While it is easy to ridicule Huxley for his seemingly ignorant carte blanche to "fish, baby, fish," it is harder to make a quantitative assessment of the impact of fishing on a stock and a reliable estimate of fisheries reference points. Fortunately, the description of fish demography developed in chapter 4 has provided us with a solid foundation. We have only to add a description of the fishing mortality to assess the impact on the size structure and recruitment of a fished stock. From there, it is fairly straightforward to calculate fisheries reference points. In this chapter, I will exploit the demographic model to make impact assessment of fishing and calculate fisheries reference points for fish stocks with asymptotic sizes of 10 g, 333 g, and 10 kg. The three asymptotic sizes span the variation in fish life histories from small and short-lived forage fish species, such as sardine or sprat; to small pelagic fish, such as herring or mackerel; to large demersal species, such as cod or saithe.

When fishing is added to the demographic model from chapter 4, the model has to be solved numerically. To complement the numerical results, I will first develop a very simplified analytical model. We consider the biomass of the fished size range of stock, $B_F(t)$. The fishing pressure is represented by the fishing mortality, μ_F. The removal of biomass is described by the equation

$$\frac{dB_F(t)}{dt} = -\mu_F B_F(t) \Leftrightarrow B_F(t) = B_F(0)e^{-\mu_F t}. \tag{5.1}$$

If no new biomass is added to the fished biomass, the fished biomass declines exponentially. The fraction of the population remaining after a year is then $B_F(1\text{yr})/B_F(0) = 1 - e^{-\mu_F}$ (assuming μ_F is measured in units of per year). A fishing mortality of 1 yr^{-1} will therefore remove $1 - e^{-1} \approx 63$ percent of the fished population per year, while a fishing mortality of 0.25 yr^{-1} will remove approximately 25 percent per year.

5.1 FISHERIES SELECTIVITY

The simple impact assessment did not specify the size range where fishing acts but considered only the change in biomass. All fishing gear, however, select fish within a certain size range. The size range affected is described by the gears' *selectivity* (fig. 5.2): gill nets preferentially select fish within a narrow size range, hooks on long lines and trawl catch anything large enough to bite the hook or avoid slipping through the mesh in the cod end, while traps, used in some fisheries for cod or crustaceans, retain only individuals small enough to enter the trap opening. The selectivity describes the relative selection of different sizes by a gear; to get the actual fishing mortality on the stock, the selectivity $\psi_F(w)$ is multiplied by the maximum fishing mortality F

$$\mu_F(w) = F\psi_F(w). \tag{5.2}$$

The gear selectivity matters on two accounts. First, the extent of the selected size range affects the cumulated impact of a given fishing mortality. Clearly, the trawl selectivity in fig. 5.2 will have a larger impact than the gill net selectivity because it affects a larger range of sizes. Second, it matters when in life the fishery

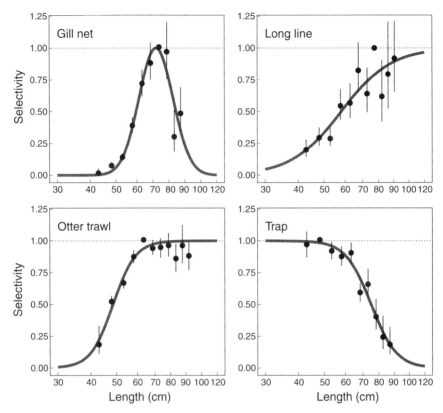

FIGURE 5.2. Relative selectivity of various fishing gear on cod, normalized to be maximally 1. Gill nets are fitted with a log-normal selection curve (eq. 5.4) and the other gears with a sigmoidal function (eq. 5.3). Data points are from Myers and Hoenig (1997).

acts—in particular, whether it selects for immature or mature individuals. Mature individuals may already have had a chance to spawn, while immature have not.

An exhaustive characterization of all possible types of selectivity is beyond my capacity here. I will mainly focus on trawl and gill net selectivities because they are the dominant types of selectivity in industrial fisheries. The selectivity of a trawl is a sigmoidal function (fig. 5.3)

$$\psi_{\text{trawl}} = (1 + (w/w_F)^{-u})^{-1}, \tag{5.3}$$

where w_F is the inflection point at the size with 50 percent retention, and $u = 3$ is a nondimensional parameter describing the sharpness of the selection around the size of 50 percent retention; $u \to \infty$ gives a "knife-edge" selectivity. Fisheries employ a mesh size suitable for the targeted species: the larger the species, the

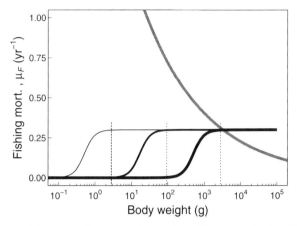

FIGURE 5.3. Trawl fishing mortality curves for species with asymptotic sizes $W_\infty = 10$, 333, and 10,000 g and a maximum fishing mortality of $F = 0.3$ yr^{-1}. Retainment of 50 percent is at $w_F = \eta_F W_\infty$, with $\eta_F = 0.05$, and the steepness is $u = 3$. The gray line is the predation mortality $\mu_p = aAw^{n-1}$ (eq. 4.7). The vertical dotted lines are at the size of 50 percent maturation—these gears target both juvenile and adult individuals.

larger the mesh size is used, and therefore the larger the value of w_F. I assume that w_F is proportional to W_∞: $w_F = \eta_F W_\infty$ with $\eta_F \approx 0.05$. Gill net selectivity is described by a log-normal function

$$\psi_{\text{gillnet}} = \exp\left[-\log^2(w/w_F)/\sigma_F^2\right], \qquad (5.4)$$

where w_F is at the maximum and $\sigma_F \approx 1.5$ characterizes the width.

5.2   IMPACT OF FISHING ON SMALL AND LARGE SPECIES

Comparing fishing mortality to the natural mortality from predation provides a first estimate at how fishing affects species with different asymptotic sizes (fig. 5.3). The predation mortality on small species dominates over the fishing mortality, while on large species the fishing mortality is comparable to the natural mortality. We therefore expect the impact of fishing with a given mortality to be smaller on small species than on large species, simply because small species have a faster life history tuned to a naturally high mortality. We must, however, not rush to this conclusion because, as shown in chapter 4, large species also have a higher density-dependent regulation that can substitute for fishing mortality. Huxley's argument for the inexhaustibility of the great fisheries was essentially based on the argument that fishing mortality was "infinitely" less than the natural mortality. His assessment that a fishing mortality of roughly 0.05 yr^{-1} was smaller than natural mortality—though not infinitely so—was fairly correct. Most developed fisheries, however, impose much larger fishing mortalities.

Before going into the full numerical simulation, I will expand the simple argument of the impact on fished biomass B_F from eq. 5.1 to consider species with different asymptotic sizes and growth rates. The fished biomass B_F will be determined by a balance between the biomass of fish growing into the fished size range J_F (mass per time) and the losses due to predation and fishing mortality. The predation mortality is described as aAw_F^{n-1} (eq. 4.7). In steady state, the flux into the fished size range equals the losses: $J_F = aAw_F^{n-1}B_F + FB_F$, where FB_F is losses to fishing. Assuming that $w_F \propto W_\infty$, B_F is proportional to

$$B_F \propto \frac{J_F}{aAW_\infty^{n-1} + F}. \tag{5.5}$$

When we consider the reduction of biomass due to fishing $B_F/B_F(F=0) - 1$, the flux into the fished range, J_F, disappears

$$\frac{B_F}{B_F(F=0)} - 1 = \frac{aAW_\infty^{n-1} + F}{aAW_\infty^{n-1}} - 1 = \frac{F}{aA}W_\infty^{1-n}. \tag{5.6}$$

This result shows two things. First, the reduction in biomass increases with asymptotic size $\propto W_\infty^{0.25}$ (for $n = 3/4$). Second, it shows how the reduction depends on the ratio between fishing mortality and the growth coefficient A: fast-growing species (higher A) tolerate a higher fishing mortality F than slow-growing species. This seems trivial, but remembering that a higher growth also implies a higher predation mortality—as predation mortality is proportional to A—this result is less obvious.

To obtain a full impact assessment, we need to combine the demographic model from chapter 4 with the fishing mortality from section 5.1. The model equations and the parameters are summarized in appendix A. In the simplest case, a stock is described by its asymptotic size W_∞, and fishing by the level of fishing mortality F and the size of 50 percent retainment by the fishing gear w_F. The result of the impact assessment is the biomass size spectrum $B(w)$, the spawning stock biomass B_{SSB}, and the recruitment relative to the maximum recruitment R/R_{max}.

The impact of fishing on the full demographic model is shown in fig. 5.4. As anticipated by the simple argument in eq. 5.6, larger species are harder hit by fishing than small species.

The reduction in spawning stock biomass and recruitment is shown in fig. 5.5a. The spawning stock biomass appears to be very sensitive to a low fishing mortality, but less sensitive to large mortalities. This is because at large mortalities, much of the large fish are completely absent (see fig. 5.4). This reduction in effective size

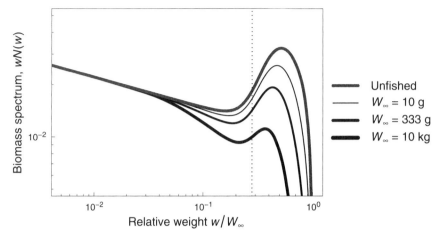

FIGURE 5.4. Biomass spectra $B(w) = wN(w)$ of three species with asymptotic size $W_\infty = 10$ g, 333 g, and 10 kg (increasing line width). The stocks are subjected to trawl-selectivity fishing with mortality $F = 0.3$ yr^{-1}, as shown in fig. 5.3. The gray line shows the unfished spectrum, which is the same for all stocks. The bump around the size at maturation at $w_m/W_\infty = \eta_m = 0.28$ (dotted vertical line) is due to the reduction in growth rate when the individuals mature and begin to allocate energy to reproduction (see also fig. 4.2).

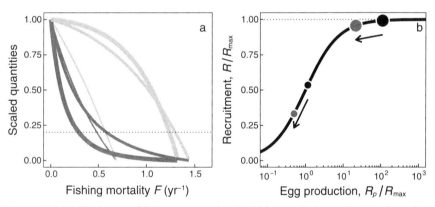

FIGURE 5.5. (a) The impact of fishing on spawning stock biomass (dark gray lines) and recruitment (light gray lines) on three stocks with asymptotic sizes 10 g, 333 g, and 10 kg (increasing line width). (b) The impact of fishing on recruitment with fishing mortality $F = 0.3$ yr^{-1} on a small species (10 g, small circles) and a large species (10 kg, large circles). The unfished recruitment is black, the fished gray.

range of fish means that a larger increase in fishing mortality is needed to reduce spawning stock by the same fraction as when the stock structure is intact.

The impact of moderate fishing is less on recruitment than on spawning stock biomass. This is because recruitment is buffered by the stock-recruitment function. Recruitment of a species that is on the flat upper part of the stock-recruitment curve in fig. 5.5b will not be affected much (unless the fishing mortality is very high), while there will be an immediate effect of fishing on a species on the rising part of the curve. As large species are on the flat part of the stock-recruitment curve, their recruitment is therefore expected to be relatively unaffected by fishing, while the recruitment of smaller species on the rising part of the stock-recruitment curve is immediately affected.

5.3   FISHERIES REFERENCE POINTS

A stock's resilience to fishing can be condensed into a set of *reference points*. A reference point is a number that indicates a characteristic level of spawning stock biomass or fishing mortality. Reference points are the pillars of harvesting rules in contemporary fisheries management: *target* reference points are management goals, and *limit* reference points are states to be avoided. Two types of reference points are needed: *biomass* reference points relate to the size of the stock, while *exploitation* reference points relate to the level of fishing mortality. For example, a stock may be considered in good shape when the spawning stock biomass is close to the target, but if the fishing mortality is above the limit exploitation reference point, the stock is overexploited. The stock will be on a trajectory that will see the spawning stock biomass dipping below the biomass limit within a few years unless the management takes actions to lower the fishing mortality. Fig. 5.6 gives an example of how reference points are used for advice in practice.

A central, almost mythical, reference point is related to the maximum sustainable yield (MSY). The MSY can refer either to the fishing mortality F_{msy} that delivers the MSY or to the spawning stock biomass B_{msy} of a stock in equilibrium and exploited with F_{msy}. MSY was hailed as a savior only to be derided as a traitor (Larkin, 1977). Now, MSY has been reinstated as the gold standard (Hilborn and Stokes, 2010); in the United States by the Magnuson-Stevens fisheries management and conservation act,[1] in the EU by the revised Common Fisheries Policy, and internationally by the United Nations Convention on Law of the Sea (UNCLOS). The MSY espouses a view on fish stocks as production systems—a

---

[1] Though the United States does not aim for the F_{msy}, the concept of optimal yield is still defined with reference to MSY.

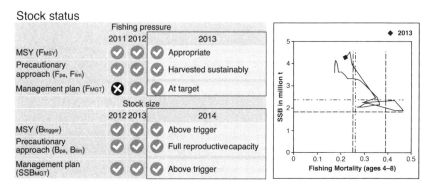

FIGURE 5.6. Example of advice produced by the International Council for Exploration of the Sea (ICES) for mackerel in the Northeast Atlantic. The advice relates the exploitation reference points (top three rows) and biomass reference points (bottom three rows) to the magnitude of the fishing mortality and the spawning stock biomass (right panel). In this case, the status is good because spawning stock biomass is well above limits and the fishing mortality is close to the F_{msy} but away from F_{lim} (vertical dashed-dotted line; ICES Advice for Northeast Mackerel, 2015).

fish stock is a factory and the goal of fisheries management is to run the factory efficiently to obtain the highest output (highest fisheries yield) with the least effort (lowest fishing mortality). While the meaning of *maximum* in MSY is evident, the meaning of *sustainable* and *yield* are less so. What is considered sustainable is to a large degree a choice of values (Quinn and Collie, 2005). Commonly quoted is the Brundtland Commission's definition of sustainable development as meeting "the needs of the present without compromising the ability of future generations to meet their own needs" (Brundtland et al., 1987). Within the context of MSY, *sustainable* is often interpreted as *sustained*—that is, that the maximum yield can be taken from the stock indefinitely, which is a fairly narrow definition of *sustainable*. *Yield* is typically interpreted as landed biomass per time, but it could just as well be stated in economic terms, such as the monetary value of landings per year or economic rent. This ambiguity in MSY is well represented in the UNCLOS definition, which relates the production of "the maximum sustainable yield, as qualified by relevant environmental and economic factors, including the economic needs of coastal fishing communities and the special requirements of developing States" (UNCLOS Article 6.13). I will use the narrow definition of *yield* as being the fisheries yield (landed or discarded) measured as biomass per time, and *sustainable* to mean that this yield can be sustained over time. This definition is predominant in contemporary fisheries management.

Fisheries yield (biomass per time) is calculated by multiplying the biomass with the fishing mortality (eq. 5.2) and integrating over the entire size range

$$Y = \int_{w_R}^{W_\infty} \mu_F(w) B(w)\, \mathrm{d}w. \qquad (5.7)$$

A related and often used measure is the "yield per recruit" $Y_R = Y/R$, with dimensions biomass per recruit per time. This measure ignores the effect of diminishing recruitment due to fishing, which we saw in fig. 5.5, and is therefore sensible only when a stock's recruitment is known to be unaffected by fishing.

The most common reference points are illustrated by plotting the state of the stock (spawning stock biomass B_{SSB} and recruitment R), and the catch from the fishery (yield and yield per recruit) as functions of the fishing mortality (fig. 5.7). Fisheries yield increases as fishing is intensified and peaks around the fishing mortality, where recruitment becomes affected by fishing, at F_{msy}. The yield per recruit is unaffected by the decline in recruitment and decreases weakly only after its maximum, at F_{max}. The F_{msy} defines the biomass reference point B_{msy} as the spawning stock biomass when the stock is exploited at F_{msy}.

While the definition of the MSY and maximum yield per recruit reference points are evident, the definition of limit reference points vary. Limits are typically defined as the point where recruitment is impaired (ICES, 2000), without a precise definition of what impairment entails. I define the limit reference points as the point where recruitment is reduced to half the maximum recruitment: $R/R_{\mathrm{max}} = 1/2$. I have also defined the reference point F_{crash} as the point where $R_0 = 1$. Further reference points may be defined, such as the "precautionary" or "management" reference points used by ICES (fig. 5.6) or the "optimal yield" reference points used in the United States. They are usually derived from the MSY and limit reference points.

Plotting the reference points versus asymptotic size summarizes how different fish species respond to fishing (fig. 5.8). The fishing mortality reference points are roughly independent of asymptotic size, until some small asymptotic size where the stocks become very sensitive to fishing. The only exception is the F_{max} reference point, which measures the maximum yield per recruit. F_{max} can be a very misleading target reference point because it suggests that small species should be fished above the level where they crash. This failure to correctly represent the state of the stock is a result of ignoring fishing effects on recruitment.

Fisheries management often determines whether a stock is "collapsed" with reference to the spawning stock biomass (Yletyinen et al., 2018). This is commonly done by comparing current spawning stock biomass to the unfished biomass B_{SSB0}. A reduction of B_{SSB} to below $0.2B_{\mathrm{SSB0}}$ is commonly used as a reference (I will use that criterion later in chapter 12). The dotted line in fig. 5.8b shows how that benchmark is somewhat arbitrary and not a good proxy of either MSY or limit reference points. In particular, large species are expected to experience more than

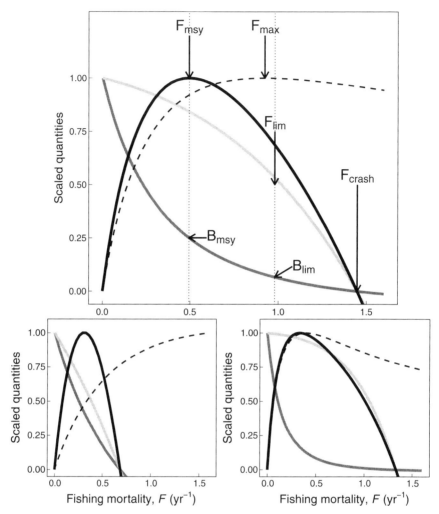

FIGURE 5.7. Demographic quantities as functions of fishing mortality: yield (black lines), yield per recruit (dashed lines), spawning stock biomass (dark gray lines), and recruitment (light gray lines), all scaled by their maximum value. The stock is fished with a trawl selectivity as in fig. 5.3. The panels represents species with different asymptotic sizes: 333 g (top), 10 g (bottom left), and 10 kg (bottom right).

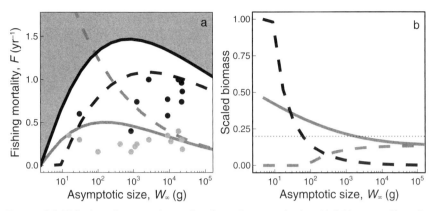

FIGURE 5.8. Fisheries reference points as functions of asymptotic size: (a) fishing mortality reference points, and (b) biomass reference points relative to unexploited spawning stock biomass. Solid black line: F_{crash}; dashed black lines: F_{lim} and B_{lim}; solid gray lines: F_{msy} and B_{msy}; dashed gray lines: F_{max} and B_{max}. Circles show F_{msy} and F_{lim} reference points for ICES stocks from Andersen and Beyer (2015).

80 percent reduction in spawning stock biomass when they are fished at F_{msy}. The practical application of the $0.2B_{SSB0}$ rule is further complicated by the difficulty in determining the unfished biomass because pristine stocks are rarely assessed.

I have shown how reference points depend on asymptotic size, but they also depend on other parameters—notably, the growth coefficient A, the physiological mortality a, and the recruitment efficiency ε_R. The dependency on the growth coefficient was derived earlier (eq. 5.6): the fishing mortality is proportional to A; a fast-growing cod tolerates a higher fishing mortality than a slow-growing redfish.

A change in the physiological mortality a represents a change in the fish community surrounding the target stock. The physiological mortality is the ratio between the level of predation mortality and the growth rate (section 4.4). An increased a indicates increased predation pressure or a lack of food. By changing a, we can therefore make a first assessment of how changes in the ecosystem affect a stock (fig. 5.9). Lowering a results in more resilient fish stocks (higher value of F_{lim}) and higher yields (higher B_{msy}) that are typically also reached with smaller efforts (smaller F_{msy}). One example of how a change in the fish community can bring about a change in a is the overfishing of large demersal stocks that has happened in fisheries in the North Sea (Daan et al., 2005) and the Northwest Atlantic (Frank et al., 2005). The depletion of these stocks reduced the predation pressure on small species—smaller a—which led to an increased biomass of smaller species. The reduction in demersal stocks therefore facilitated the huge productivities of forage fish stocks, such as the sand eel in the North Sea. As several demersal stocks in the North Sea are currently on track for a recovery, we can expect the

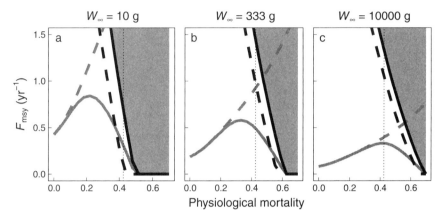

FIGURE 5.9. Fishing mortality reference points as a function of the physiological mortality a. Lines as in fig. 5.8; the dotted vertical line is the default value of a.

converse to occur in the coming years: as large species recover and increase in biomass, small species will experience increased mortalities, and thus increased a, with a concomitant need for reassessment of reference points and reduction in the fishing mortality on these species (van Gemert and Andersen, 2018a). I will make a more detailed evaluation of such ecosystem effects in chapter 12.

The recruitment efficiency ε_R represents survival of larvae until the size at recruitment. We can expect ε_R to vary from year to year due to the stochastic nature of larvae survival. Studying the effect of fluctuations in detail requires a dynamic model, which will not be developed until chapter 7. Nevertheless, we can still proceed with qualitative arguments. The impact of the variability in ε_R on a stocks' reference points varies with asymptotic size (fig. 5.10): large species are expected to be almost unaffected by variations in recruitment efficiency—at least as long as recruitment does not fail entirely—while the F_{msy} reference point of smaller species is very sensitive to changes in recruitment efficiency. These differences between species reference points again reflect differences in the strength of density dependence between small and large species (section 4.2 and fig. 5.5): small species have little density-dependent regulation and are thus sensitive to changes in the environment, while the large density-dependent regulation in large species buffers the environmental variability. Consider the North Sea sand eel as an example of a small species with an asymptotic weight of about 20 g. The North Sea stock supports a huge fishery that is subjected to large annual fluctuations in catch due to variable recruitment of the stock. Because the stock is sensitive to recruitment, management must estimate recruitment every year and use this assessment as a basis for the annual quota.

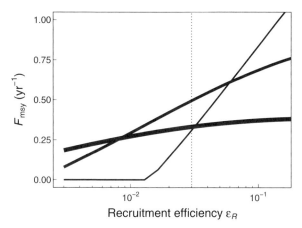

FIGURE 5.10. F_{msy} as a function of ε_R for species with asymptotic size 10 g, 333 g, and 10 kg (thin to thick lines). The dotted line indicates the value of ε_R used in fig. 5.8.

5.4   WHICH GEAR SELECTIVITY MAXIMIZES YIELD?

So far I have focused on trawl selectivity with a 50 percent selection at $0.05W_\infty$ as a reasonable representation of large industrial fisheries. The size selectivity of a gear, however, has a huge influence on the fishing mortality reference points. For example, a gear that selects for a narrow size range requires a very large fishing mortality to extract the same yield as a gear that selects for a wide size range. For yield-maximizing fisheries management the question is: which size range of fish produces the highest yield? We found a partial answer to that question in fig. 4.3: if we ignore recruitment, the biomass of a cohort peaks around size of maturation, and this is the size where we should harvest the entire cohort. While this result qualitatively anticipates the coming results, it ignores that the stock should also be allowed to reproduce to maintain recruitment. To examine that aspect, I calculate the yield for different size selectivities with three different gears: idealized "knife edge" selection, trawl, and gill net (fig. 5.11).

All gears show a similar pattern: yield is maximized if the fishery selects mature individuals. How much larger than size at maturation depends on the gears. Knife edge selectivity maximizes by selecting close to maturation, while the imperfectly selecting gear should focus on larger sizes. This result is very similar to the idealized argument that the cohort should be harvested completely at the peak of its biomass, and it is also well known; Beverton and Holt calculated it and constructed the cardboard model seen in the background of fig. 1.1. The curves of optimal size selectivity show another important result: there is a wide size range where the yield is close to the maximum. This is most evident for the trawl selectivity that retains

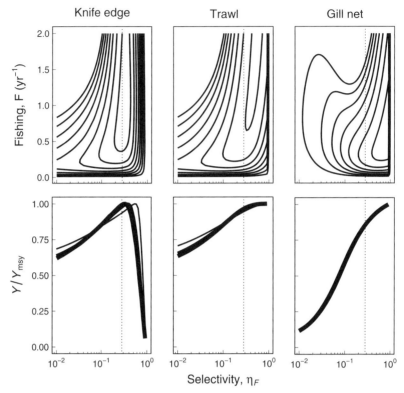

FIGURE 5.11. How maximum yield changes with the size at selection for different types of gear. Top: yield as a function of the 50 percent retainment relative to asymptotic size, $\eta_F = w_F/W_\infty$, and fishing mortality F for a species with $W_\infty = 10$ kg; bottom: the maximum yield as a function of the selectivity for species with $W_\infty = 10$ g, 333 g, and 10 kg (increasing line width). The vertical dotted lines are at the size of 50 percent maturation.

between 80 and 100 percent of the maximum yield over a size range of a factor of 10. A precise size selection is therefore not crucial to obtain (almost) maximum yield.

5.5   SUMMARY

We have now formulated a complete theoretical framework that can be applied to make ecological impact assessments of fishing a single stock. The framework can make impact assessments broadly on all fish species solely by varying the asymptotic size and using the "default" life-history parameters in table A.2, or it can be applied to a specific stock using parameters specific to that stock, or even an updated growth equation that fits that particular stock better. I have shown how the framework can be used to estimate reference points and thereby reveal how the resilience to fishing depends on the parameters.

The generality of the framework makes it applicable to other questions. An obvious example is to examine the importance of the "BOFFs" (the big old fecund females) for the reproductive potential and the recruitment of a stock. The BOFFs have received a lot of attention because they have a high egg production and they also seem to produce eggs and larvae with a higher survival probability (higher ε_R; Hixon et al., 2013; Barneche et al., 2018). As most fisheries predominantly target the largest fish in a stock, it is relevant to consider whether the potentially important BOFFs should be particularly protected. An application of the size-based framework provides a weighing between the higher productive potential of the BOFFs with their lower abundance, and thereby assesses under which conditions (and possibly which stocks) protection of the BOFFs should be prioritized by management. Such an application was done by Calduch-Verdiell et al. (2014), who showed that the BOFFs contribute only a small part of the total spawning potential of a fished stock, even when the higher survival probability of their eggs are considered. Efforts to protect the BOFFs are therefore unlikely to result in notable protection or resilience of the fished stock.

I have calculated reference points based on MSY and recruitment impairment. These reference points describe how demography responds to exploitation. For management purposes, other aspects may be woven into the reference points. A variant of the MSY target is the maximum economic yield (MEY) (Gordon, 1954; Grafton et al., 2010). The MEY accounts for two economic aspects: that larger fish are typically more valuable than smaller ones, and that fishing effort has a cost, typically proportional to the fishing mortality, the landings, or a combination. Both aspects result in a maximum yield at a lower fishing mortality than the MSY and a correspondingly higher biomass. A reduction in the target reference point below F_{msy} might not even reduce yield much below the MSY because the biomass yield versus fishing mortality curves are typically fairly flat around the optimum (see fig. 5.11). Further, reducing the target exploitation reference point also ensures against an unobserved fluctuation in recruitment compromising the stock. The suggestion of a "pretty good yield" by Ray Hilborn is born out of such practical considerations (Hilborn, 2010), as is the concept of "optimal yield" in the United States. Both cases argue that the benefits of adopting a target exploitation lower than MSY compensates for the loss in biomass yield.

While a lot of scientific effort is invested into assessing the state of a fish stock—the current fishing mortality and stock size—less effort is invested in estimating the reference points. This is unfortunate, because management decisions are based not on the absolute values of the state of the stock but on the values relative to reference points—essentially, about whether the state is above or below the reference points. The problem is of course that determining the reference points is difficult because they depend on the life-history parameters, and in particular on growth and natural mortality. The growth parameter A is stock specific and fairly easily estimated from von Bertalanffy growth parameters (box 3.2). It is harder to estimate the physiological mortality a and the recruitment efficiency ε_R that together describe the environment. Further, the environment is rarely stable over longer periods, so even when averaging out annual recruitment fluctuations, these parameters are continuously changing. The productivity of a fish stock is therefore not, as often perceived, a property of a species or a stock, but shifts with the ever-changing biotic and physical environment. Consequently, reference points depend on the environmental context and should be reevaluated frequently.

The methodology developed in this chapter can also be used as a basis for "data-poor" stock assessments. We tend to think of fisheries management as being done by advanced industrialized nations to manage highly productive iconic stocks, such as anchovies, herring, cod, tuna, salmon, and so on. The economic and cultural importance of these stocks justify large expenses for sampling and big investments in scientific effort to obtain the best possible assessments and impact assessments of the fishery. However, the overwhelming majority of fished stocks in the world have not been offered the same lavish attention, and consequently management operates partly blindfolded because of lack of information—if they are managed at all. Such data-poor situations are typical in the developing world, but they are actually also common for many by-catch species in industrialized fisheries. In the latter case, the target stock itself might be managed with all the bells and whistles afforded by rich nations, but the ecology of the by-catch species might be largely unknown. Development of fisheries advice for data-poor stocks requires methods that can make the most of the little available information, and bring in information from other similar stocks for support. This way of making the data-rich stocks assist the data-poor stocks has been called the "Robin Hood" approach (Punt et al., 2011). The trait-based approach provides a formal way of implementing the Robin Hood approach. In even the most data-poor situation, we have an idea of the maximum size of the fish, which can be used as a decent first estimate for the asymptotic size. Knowing the growth rate coefficient is harder, because it requires that the ages of fish of different sizes are established, which is rarely the case. Fortunately, as we saw earlier, the growth rate parameter matters only a little for population-level measures. Knowing the asymptotic size, a stock assessment of fishing mortality can be developed on the basis of the sizes of fish in the catch, and reference points can be estimated. Both measures, the estimated fishing mortality and the fisheries reference points, are almost proportional to the value of the growth rate coefficient used. Therefore, not knowing A

might be problematic. However, when we form the ratio of the two, F/F_{msy}, the importance of A cancels out (Kokkalis et al., 2015). We can therefore make reliable statements about whether a stock is overexploited ($F/F_{\mathrm{msy}} > 1$) or not, even when the growth-rate parameter is unknown. This method has been validated on data-rich cod stocks with great promise (Kokkalis et al., 2017).

The most striking revelation in this chapter is the remarkable resilence of fish stocks to exploitation. The calculations show that about a quarter of a stock can be removed each year, even of long-lived species, without compromising production. On top of that, the average species tolerates a fishing mortality in excess of $1 \ \mathrm{yr}^{-1}$ before it collapses entirely. Among long-lived organisms, this is a remarkable resilience. What is also remarkable is that small-bodied species are less resilient than large-bodies species. This result runs against predictions of metabolic arguments. A metabolic argument essentially states that all rates scale with body size to the $-1/4$ exponent—that is, to $W_\infty^{-0.25}$ (Brown et al., 2004; see box 4.1). The calculation of reference points showed that only the F_{max} reference point followed the metabolic scaling, all other fisheries mortality reference points do not follow a $W_\infty^{-0.25}$ scaling. This is because large-bodied fish stocks have a big buffer of density-dependent recruitment that can be exchanged for fishing mortality. It is therefore not the small-bodied species that are particularly sensitive to fishing, it is the large-bodied species that are exceptionally resilient. An important exception is the large sharks and rays, which are very sensitive to fishing—I'll return to the sharks in chapter 8.

CHAPTER SIX

Fisheries-Induced Evolution

In the previous chapter, I considered the demographic effects of fishing—how fishing changes the size structure and the recruitment. Yet, fishing leaves deeper impacts than just demographic changes. Most fisheries are size selective, and that selectivity will impose a Darwinian evolution on the fish stock. A typical fishery targets the largest individuals. The selection of large individuals means that individuals with certain traits are at larger risk of being caught than other individuals. For example, slower growing individuals will live longer than faster growing individuals before they reach the size where they are targeted by fishing. If the longer life also translates into more spawning events than the faster growing individuals, then slower growing individuals will be favored in the next generation. Consequently, the evolutionary "selection response" will be slower average growth in the population. How the selection by fishing changes the genetic makeup of the fished population is similar to how humans have improved crops or livestock for millennia. The difference is that the selection responses of fishing are unintended, and they do not necessarily improve the stocks' productivity.

The evolutionary side effects of fishing were first thoroughly explored by Richard Law and David Grey's theoretical study on the Barents Sea cod (Law and Grey, 1989). The Barents Sea cod has supported a coastal fishery in the vicinity of the Lofoten Islands for at least a thousand years. We know that the Scandinavian Vikings exported dried cod,[1] and it has even been proposed that the availability of a nutritious and long-lasting food was the key element that made their long sea voyages possible (Kurlansky, 1998). The historic fishery off the Lofoten Islands mainly targeted mature cod when they migrated in from the North Atlantic to spawn near the coast. Such a "spawner fishery" imposes a selection that favors late-maturing individuals over early-maturing individuals. Late-maturing individuals will be very big once they enter the spawning ground, where the risk of being

---

[1] "Thorolf had a large ship...; he freighted it with dried fish and hides, and ermine and gray furs too in abundance, and other peltry such as he had gotten from the fell; it was a most valuable cargo. This ship he bade sail westwards for England.... There they found a good market, laded the ship with wheat and honey and wine and clothes, and sailing back in autumn with a fair wind came to Hordaland." Egil's Saga from 850 A.D.

caught is high. Conversely, younger spawners will be small when they spawn and will therefore get a lower benefit (fewer eggs) from exposing themselves to the risk of entering the spawning grounds. The selection pressure from the spawner fishery has therefore been to develop later maturation. In the 1930s, the offshore fishery on feeding grounds was developed with trawlers. This "feeder fishery" exposed the stock to a new selection pattern by also targeting juveniles. The targeting of juveniles completely reversed the selection pressure on size at maturation. Now the feeding grounds were no longer safe havens, and those individuals that delayed spawning risked being caught before they even made it to the spawning grounds. The selection pressure from the new feeder fishery was therefore to develop earlier, and not later, spawning. Law and Grey noted that age at maturation in the Barents Sea stock had declined substantially, from about 9 years to 7 years, just 20 years after the commencement of exploitation by modern trawlers. To support their suggestion of fisheries-induced evolution, they developed a theory that showed how the optimal age at maturation would indeed decrease following the change in exploitation from a spawner fishery to a feeder fishery. While the theory formalized the simple argument that I developed earlier for why fishing creates a selection response on size at maturation, it did not address a crucial question: How fast are the selection responses? With fishing mortalities often reaching 1 year^{-1}(corresponding to an annual removal of 63 percent of the stock, p. 84), the selection pressures are substantial and one would expect rather fast selection responses.

Law and Grey's work stirred the interest of David Conover and Stephan Munch at Stony Brook. They designed an experiment that could answer how fast fishing could change the genetic makeup of a fish population. The difficulty with such an experiment is that it needs to run over many generations, which would take decades for most fish populations. To make the experiment feasible, they chose to work on Atlantic silversides because they have a generation time of just one year. For five years, they subjected populations of silversides in the laboratory to three harvesting regimes: preferential harvest of the fastest growers, preferential harvest of the slower growers, and random harvest as a control treatment. The results shown in fig. 6.1 are very clear: five generations of harvesting the fastest growers created a dominance of slower growing individuals and vice versa (Conover and Munch, 2002). The selection response—the change in the mean value of a trait (the body size at 185 days) was around 8 percent per generation. Conover and Munch continued their experiments on the silversides to explore whether the evolutionary changes brought about by the harvesting would be reversed if harvesting was stopped. Those experiments confirmed a troubling hypothesis from Law and Grey's model: the changes can be reversed, but at a much slower rate (fig. 6.1b; Conover et al., 2009). Taken together, the experiments

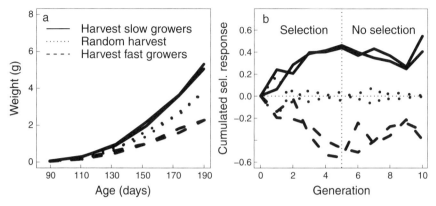

FIGURE 6.1. Changes in mean length of adults observed in experiments on Atlantic silversides. During the first five generations, the population is exposed to three different harvesting regimes: harvesting slow-growing individuals (solid line), harvesting fast growing individuals (dashed line), or random harvesting (dotted line). After the fifth year, the selection is removed and the population is left alone. (a) Weight-at-age after the first five generations. (b) Cumulated selection response as the change in body mass per average body mass (3.7 g) during the selection (first five generations) and after selection is terminated (last five generations). Redrawn from Conover and Munch (2002) and Conover et al. (2009).

demonstrated how selective harvesting can bring about almost irreversible changes within a few generations.

The experiments with silversides had one deficiency: they did not reflect the selection pressures of real fisheries. By selectively removing fast- or slow-growing individuals, Conover and Munch applied a selection pressure directly on a trait, in their case the growth rate. Yet, as we saw in fig. 5.2, fishing gear does not select directly on a trait, it selects on the size of individuals, and predominantly selects the largest individuals. Conover and Munch also harvested the largest individuals, but they did so at a specific date, and thereby they selected directly on the growth rate. Real fisheries rarely happens on just one date but occurs continuously. Therefore, the slower growing individuals will also be exposed to size-selective fishing, only a little later than the fast-growing individuals. Consequently, a size-selective fishery will catch both slow- and fast-growing individuals, albeit with a preference for the faster growing individuals. Fishing, therefore, does not select directly on a trait, such as growth rate, but does so only indirectly through the selection on body size. The selection responses observed by Conover and Munch are therefore exaggerated. While the experiments clearly demonstrate the potential for fishing to induce an evolutionary response, and that changes are difficult to reverse, they do not answer the crucial question about the rate of the evolutionary changes in real fisheries.

The evidence that could finally settle the question about the rate of fisheries-induced evolution would be a direct observation of a change in a trait during a period of heavy fishing. Law and Grey did point to the rapid changes in age at maturation in Barents Sea cod as an indication of evolutionary changes. However, as they were well aware, such changes need not be evidence of genetic changes in the population. Direct observations of changes in traits indicate only phenotypic changes, and not necessarily genotypic changes. Changes in maturation or growth might be due to a shifting environment or simply density-dependent effects brought about by the large changes in population size associated with fishing. Northern cod, off southern Labrador and eastern Newfoundland, is another large cod stock that has supported fishing for centuries. As with the Barents Sea cod, Northern cod has also shown phenotypic changes in age at maturation. To separate phenotypic changes from genotypic changes, Olsen et al. (2004) looked for changes in the *reaction norm* of a trait, rather than at the trait itself. The reaction norm describes how the phenotypic expression of a trait co-varies with the environment or with another trait. Examples of reaction norms are how growth rate co-varies with temperature, or how age at maturation co-varies with growth rate. Changes in the expression of a trait along the reaction norm are indicative of phenotypic plasticity, while changes in the reaction norm itself would be strongly indicative of a genotypic change. Olsen and co-workers demonstrated that the reaction norm had indeed shifted during the period of heavy fishing with a rate of around 2 percent year^{-1} (1 cm/year). With a generation time of around five years, this rate is even faster than observed in the experiments by Conover and Munch, despite the selection in the fishery being only indirect.

While the changes in reaction norms is a useful way to assess rates of evolutionary changes, the method is not foolproof. For example, the reaction norms might harbor a hidden dimension that is not revealed in the two-dimensional reaction norms used by Olsen and co-workers. A given trait—for instance, the phenotypic expression of age at maturation—might depend on an environmental variable in addition to growth rate. If that was the case, the observed change in the two-dimensional reaction norm could also be due to changes in this unobserved phenotypic co-variate. Nevertheless, the combined evidence from many analyses of fisheries-induced evolution (Jørgensen et al., 2007), and, as we shall see, their qualitative agreement with theory, lends support to the use of changes in reaction norms to explore the rates of fisheries induced evolution. Despite this, there is still uncertainty about the actual rates of fisheries-induced evolution (Audzijonyte et al., 2013).

The preceding description of the development of the thinking about fisheries-induced evolution is not quite accurate. The idea that fishing could induce evolutionary changes was known in general terms early in the 1980s, and Olsen and co-workers were not the first to focus on reaction norms—Adriaan Rijnsdorp estimated changes in reaction norms of North Sea plaice in 1993 (Rijnsdorp, 1993). See Law (2000) for a more complete and very readable introduction to the topic. Nevertheless, my earlier description fairly well represents the development in the mainstream thinking about the topic. Anyway, by the mid-2000s several empirical investigations had demonstrated how fishing would be a very plausible explanation for observed changes in life-history traits (Jørgensen et al., 2007). The evidence for the evolutionary effects of fishing calls for fisheries management to account for them: What are the expected rates of change? And what are their impact on the productivity of fish stocks? In this chapter, I will make such evolutionary impact assessments of fishing by combining the size-based theory developed in chapters 3 and 4 with classic quantitative genetics.

6.1   WHICH SELECTION RESPONSES DO WE EXPECT?

A size-selective fishery will impose a selection pressure on all those traits that influence fish growth and reproduction. Those traits are the ones that enter into the growth equation (eq. 3.16)

$$g(w) = Aw^n - \psi_m(w/w_m)kw, \tag{6.1}$$

with size at maturation w_m, growth rate coefficient A, and investment into reproduction k. For a population in an evolutionary equilibrium, the traits are fixed, and the variation in the phenotypic expression of trait values between individuals is caused environmental variation, by random evolutionary drift inherent in neutral evolution, local co-adaptations, frequency-dependent selection, and so on. Fishing subtly shifts the evolutionary balance, and individuals with trait values that were formerly disadvantageous will be favored. If those trait values are inherited by the next generation, the genetic makeup of the next generation will be shifted in the direction of the more favorable trait values. The core of an evolutionary calculation is laid down in the trade-offs quantifying the cost and benefits of the traits. In the demography framework developed in chapters 3 and 4, the trade-offs associated with the three traits are as follows.

Size at Maturation, w_m

The benefit of earlier maturation is a higher probability of survival to maturity, and thereby a higher likelihood of spawning at least once. However, earlier maturation also means that the fish is smaller when it reproduces. As the reproductive investment kw is proportional to body size (eq. 3.14), the reproductive output will be smaller. The cost of earlier maturation is therefore that fewer eggs are produced in each spawning event. Further, as growth slows down, the individual forgoes some of the possibility of lowering natural mortality by growing larger. Therefore, the adult survival will also be lower when maturity happens earlier.

Growth Rate Coefficient, A

A higher value of A means faster growth. The benefit of faster growth appear to be obvious: a shorter time to maturation means a higher survival to maturation. However, the assumption that faster growth is directly correlated with higher predation mortality (section 4.4) offsets that benefit. In an unfished population, the trade-off with growth is therefore assumed to be neutral because survival to maturation is independent of growth rate (eq. 4.31). However, as fishing mortality is independent of growth rate, faster growth will indeed result in higher survival to maturation, if fishing also targets juveniles.

Reproductive Investment, k

I have referred to A as the *growth coefficient* because it determines the growth rate. However, as we saw with the von Bertalanffy growth equation in section 3.1, A really scales the acquisition of resources, and a higher value of A therefore also means that more resources are available to invest in reproduction. How the available energy is divided between growth and reproduction in adults is determined by the reproductive investment, k. The benefit of increased investment in k is obviously a higher reproductive output. The downside is lower adult growth rates and smaller maximum size (eq. 3.17), which means higher natural mortality, owing to the higher mortality of smaller individuals.

A fishery that captures only mature fish, regardless of their size, will select for slower growth rates and later maturation, while a fishery that also captures juveniles makes it beneficial to mature early and grow fast. We can pose three cases that represent the major variation in the selection patterns by fisheries and hypothesize which life-history strategy is most successful in each case:

1. *A trawl selection pattern that targets both juveniles and adults.* The successful fish maximizes the likelihood that it can reproduce before it is caught. It does so by maturing early, by growing fast across the fished range to reach maturation, and by investing heavily in reproduction once it is mature.

2. *A selection pattern that targets only large fish.* The successful fish avoids becoming large by a slow growth rate, earlier maturation, and higher investment in reproduction.

3. *A spawner fishery targeting mature fish on the spawning grounds, irre-spective of their size*. The successful fish is big and fecund once it enters the dangerous spawning grounds because it delayed maturation as much as possible.

In conclusion: It is complicated. The subtle interaction between the entan-gled trade-offs of the three traits, combined with the different selection pressures imposed by typical fisheries makes even qualitative predictions of selection responses difficult. It is therefore hard to generalize observations of evolutionary changes made on one population to other populations. Even generalizing between different populations of the same species is hard, because the selection responses depend on the fishing pattern. It is therefore necessary to make evolutionary impact assessments on a stock-by-stock basis.

6.2   QUANTITATIVE GENETICS

The selection responses can be calculated with quantitative genetics. Quantitive genetics deals with traits that vary continuously, such as the size of maturation, the growth rate, or the investment in reproduction. Not all individuals in a population have the same phenotypic expression of traits, but the trait values are distributed around a mean value. This variation is a reflection of genotypic differences and environmental drivers of the phenotypic expression of genotypes. Different phe-notypes will have different fitness, but in an evolutionary equilibrium, the traits at the peak of the distribution will have a higher fitness than the traits at the fringes of the distribution. If the selection pressure on the population changes—for instance, by fishing—the trait values at the peak of the distribution may no longer have the largest fitness. In the example shown in fig. 6.2, the trait values to the right of the distribution have higher fitness than the trait values to the left. If genes are passed on directly to the next generation, the distribution of trait values will therefore be shifted to the right. The change in the mean value that would occur during one generation is called the *selection differential S_θ* of the trait θ.

Trait values are not passed directly from one generation to the next. The value of a quantitative trait is the result of the combination of a large number of alleles in the genome. Reproduction results in offspring with a new set of alleles that are a recombination of the parents' alleles. On average, offspring will have quantitative traits similar to the parents' traits, though there is a large variation. For exam-ple, my wife and I are fairly short, while all our children are substantially taller than us. The degree of similarity between the phenotypic expression of parent and offspring trait values is measured by the heritability, h^2 (the heritability should not be confused with the coefficient for maximum consumption h). A heritability

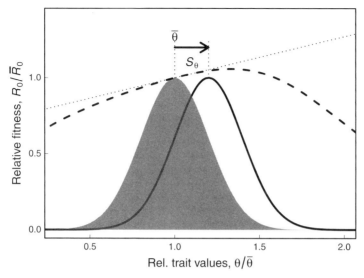

FIGURE 6.2. The principles of a quantitative genetic calculation of selection responses. The gray region shows distribution of trait values θ around the mean $\overline{\theta}$. The dashed line shows the fitness of individuals. The dotted line is the gradient of the fitness that is used in the approximation of the selection response. In this example, individuals with higher trait values are more fit than individuals at the mean of the distribution, and the selection response will evolve the distribution towards higher trait values (black line).

BOX 6.1

QUANTITATIVE GENETICS

Assume a distribution of phenotypes $p(\theta)$ characterized by a mean value $\overline{\theta}$ and a standard deviation σ_θ. With perfect copying of traits, each trait value will be copied to the next generation, and the new distribution $p^+(\theta)$ is changed by the fitness R_0 after one generation

$$p^+(\theta) = \mathscr{P}[R_0(\theta)p(\theta)], \qquad (6.2)$$

where the function $\mathscr{P}[p(x)] = p(x) / \int p(x)\, dx$ ensures that the integral of p^+ is one. We can calculate the mean value of p^+ as

$$\overline{p^+} = \int \theta p^+(\theta)\, d\theta, \qquad (6.3)$$

and the selection differential as the change in the mean value:

$$S_\theta = \overline{p^+} - \overline{\theta}. \qquad (6.4)$$

(*continued*)

(Box 6.1 *continued*)

We now expand $R_0(\theta)$ to first order around the mean: $R_0(\theta) \approx R_0(\bar{\theta}) + (\theta - \bar{\theta})R_0'$, where $R_0' = dR_0(\theta)/d\theta$ is the derivative of R_0 evaluated at $\theta = \bar{\theta}$. Inserting that expansion in eq. 6.2 gives

$$S_\theta \approx \frac{(R_0(\bar{\theta}) - \bar{\theta}R_0')\int \theta p(\theta)\,d\theta + R_0'\int \theta^2 p(\theta)\,d\theta}{(R_0(\bar{\theta}) - \bar{\theta}R_0')\int p(\theta)\,d\theta + R_0'\int \theta p(\theta)\,d\theta} - \bar{\theta}. \tag{6.5}$$

Using now that $\int p(\theta)\,d\theta = 1$, $\int \theta p(\theta)\,d\theta = \bar{\theta}$ and $\int \theta^2 p(\theta)\,d\theta = \bar{\theta}^2 + \sigma_\theta^2$, we get the simple expression

$$S_\theta \approx \sigma_\theta^2 \frac{R_0'}{R_0(\bar{\theta})}. \tag{6.6}$$

---

of 1 means that the trait is passed directly from parents to offspring, and a heritability of 0 means that the trait is completely random. It is difficult to obtain the exact value of the heritability of a given trait. In the experiments on silversides by Conover and Munch (2002), the heritability was estimated to be 0.2, and I will use that value here. Quantitative genetics estimates the change in the distribution of phenotypic traits in the population as a weighed mean between the original distribution, weighted by $1 - h^2$, and the distribution that would result if the traits were passed directly to the offspring, weighted by a factor of h^2. The actual change that occurs when the heritability is accounted for is the *selection response*, $\Delta\theta$. The selection response is then given by the *Breeder's equation*

$$\Delta\theta = h^2 S_\theta. \tag{6.7}$$

We can simplify the calculation of the selection response by assuming that the lifetime reproductive output is a linear function of the trait. As indicated with the dotted line in fig. 6.2, this is a fair assumption, in particular when the selection response is small. This assumption results in a very simple approximation of the selection response (box 6.1):

$$\Delta\theta \approx h^2 \sigma_\theta^2 \frac{1}{R_0} \frac{dR_0}{d\theta}\bigg|_{\theta=\bar{\theta}}, \tag{6.8}$$

where the fitness R_0 is the recruits per recruit (p. 73) and σ_θ is the standard deviation of the trait in the population. The derivative of the fitness R_0 is evaluated around the mean value of the trait $\bar{\theta}$. The approximation in eq. 6.8 makes sense: the selection response is faster the higher the heritability (larger h^2), the wider the distribution of phenotypes in the population (larger σ_θ), and the stronger the change in R_0 as a function of the trait. Eq. 6.8 is a recipe to obtain selection responses: by inserting our earlier expressions for the fitness from eq. 4.39, we directly get the selection response. The recipe is not quite complete yet, though—there is still one more quirk left to deal with.

In a natural population unaffected by fishing, the selection response is expected to be zero on average. This means that the population is in an evolutionarily equilibrium prior to the commencement of fishing. However, for the model developed here (and most other ones), that will not be the case. The model will not obtain a reasonable evolutionary equilibrium because it does not accurately reflect all selective drivers that act on real populations. One way to obtain an evolutionary equilibrium is by parameter tuning. If the model is set up for a particular population, one can adjust the parameters, within reasonable ranges, until the population is at an evolutionary equilibrium. I will not do that here, mainly because of the difficulty of obtaining an evolutionary equilibrium for several traits and for populations with different asymptotic sizes with the same set of parameters. To correct for the nonzero selection response, I instead calculate the relative selection differential as the difference between the selection responses with and without fishing (Andersen and Brander, 2009)

$$\Delta\theta_{\mathrm{rel}} = \Delta\theta(F \neq 0) - \Delta\theta(F = 0). \tag{6.9}$$

I also assume that the standard deviations of the traits are proportional to the mean value: $\sigma_\theta = c_{\mathrm{cv}}\overline{\theta}$, where the coefficient of variation is approximately $c_{\mathrm{cv}} = 0.2$. I further normalize the relative selection response with the trait itself to get a relative measure of the selection response. Last, to obtain a result in terms of absolute time (that is, per year), I divide the selection response by the generation time, estimated as the age of maturation t_{mat} (eq. 3.25). Taken together, the relative specific selection response becomes

$$\Delta\mathrm{rs} = \frac{\Delta\theta\mathrm{rel}}{\overline{\theta}t_{\mathrm{mat}}}. \tag{6.10}$$

The relative specific selection response estimates the change in a trait in units of the trait itself and per year. Calculating the selection responses in units of Darwins is described in box 6.2.

6.3   EVOLUTIONARY IMPACT ASSESSMENT OF FISHING

A complete evolutionary impact assessment evaluates all evolutionary consequences of fishing: trait changes, productivity changes, economic impacts, and alternative fishing scenarios (Laugen et al., 2014). Here, I focus on the direct effects on the affected traits for the three fishing scenarios from page 105: fishing

BOX 6.2

UNITS OF SELECTION RESPONSES

Selection responses, as calculated by eq. 6.8, are given in units of the trait per generation. This unit makes it difficult to compare the rates of evolution of different traits, as
each will be scaled with units of the trait itself. For example, as sizes of maturations
are often on the order of 1,000 grams, the rates of evolution on size at maturation will
appear faster than rates of evolution of reproductive investment, which is typically on
the order of 0.1 yr^{-1}. It is therefore useful to show the selection response relative to
the mean trait value—that is, $R/\bar{\theta}$. Further, to compare selection responses between
species with different asymptotic size I also scale with the expected generation time.
Getting accurate generation times in structured populations is difficult, but the age at
first maturation, as given by eq. 3.25, is a good approximation. The specific selection
response is then

$$\Delta\theta_{rs} = \Delta\theta/(\bar{\theta}t_{mat}). \tag{6.11}$$

An often-used unit of selection responses is *Darwins*. A Darwin is the logarithmic
change in a trait θ over a time period Δt, measured in millions of years: $\Delta\theta_{Darwins} =$
$(\ln(\theta(t+\Delta t)) - \ln(\theta(t)))/\Delta t$. Because the time scales considered here are short,
we can approximate the change in the trait with a linear expansion: $\theta(t+\Delta t) \approx \bar{\theta} +$
$\Delta\theta\Delta t$. Then the selection response as measured in Darwins can be approximated as

$$\Delta\theta_{Darwins} \approx (\ln(\bar{\theta} + \Delta\theta\Delta t/t_{mat}) - \ln(\bar{\theta}))/\Delta t \approx 1 + \frac{\Delta\theta}{\bar{\theta}}\frac{\Delta t}{t_{mat}}, \tag{6.12}$$

where Δt is measured in millions of years. The preceding relation shows that the
selection response in units of Darwins is the scaled selection response with age at
maturation measured in years^{-1} multiplied by 1 million and added 1.

on both juveniles and adults, fishing only on large individuals, and fishing only on
mature individuals.

Selection responses are fairly simple to calculate using the approximation in
eq. 6.8. The only thing we need is the fitness for which we can use our recipe for
recruits per recruit, R_0 (eq. 4.39) from chapter 4. I will use the life-history parameters established earlier (table A.2) and sweep over asymptotic sizes to explore the
evolutionary response of different-size species.

First, consider a fishing pattern corresponding to case I—that is, a trawl selectivity catching both juvenile and adult individuals (fig. 5.3a). The calculated directions of evolution largely confirm our qualitative predictions: fishing decreases
size at maturation and leads to faster growth rates, and higher investments in

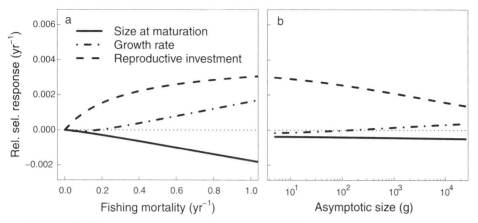

FIGURE 6.3. Selection responses for case I type selection, fishing both juveniles and mature individuals with a trawl-like selection. (a) The selection responses as a function of fishing mortality for a species with asymptotic size $W_\infty = 2$ kg. (b) Selection responses as a function of asymptotic size for a fishing mortality of $F = 0.3$ yr^{-1}. In both cases, the selectivity is a trawl with 50 percent selection at $0.05W_\infty$ (fig. 5.3).

reproductive output (fig. 6.3). Trawl fishing therefore favors a fast life history with shorter generation times.

If the mesh size of the trawl is increased, such that only large fish are targeted, the selection response on growth is reversed (case II; fig. 6.4). It is now advantageous to grow slower, because it makes it possible to spawn for a longer period. The selection responses on size at maturation and investment in reproduction are the same as in case I.

Last, consider a fishery that also includes a component of a spawner fishery (case III from page 105), which targets only mature individuals. As expected, the addition of a spawner fishery changes the selection response drastically (fig. 6.5). If the spawner fishery is dominating over the feeder fishery, the fishery leads to delayed maturation and slower growth. A spawner fishery therefore favors a slow life history.

The predicted rates of evolution in all three cases are typically less than 0.5 percent per year for the investment into reproduction, and on the order of 0.1 percent for maturation size and growth rate. This is remarkably slow when considering the strength of selection—a fishing mortality of 0.3 yr^{-1} means that 25 percent of a population is removed every year, which is a substantial selection pressure. The small magnitude of these rates contradicts the selection experiments by Conover and Munch (2002), who found that selection reduced size at age by a factor of 2 in just five years. These experiments had direct selection: they selectively removed fast-growing individuals and retained slow-growing individuals.

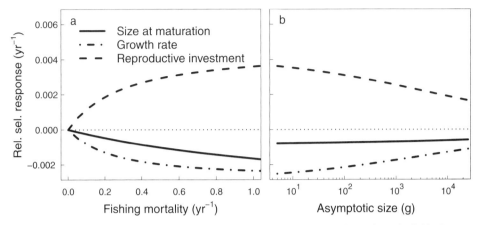

FIGURE 6.4. Selection responses for case II type selection, fishing only on large individuals with a trawl-like selection. Parameters as in fig. 6.3, but with a selection that targets only large individuals, with a 50 percent selection at $0.5W_\infty$ (fig. 5.3).

The selection by the size- or maturation-selective fishery is indirect: it selects only by size, not on the traits themselves. Therefore, the predicted selection responses are much slower than one would intuitively predict based on the expectations from direct selection experiments.

The largest selection responses are on the investment in reproduction. Investing in reproduction has a cost in terms of slower growth. Investing in growth is an investment in obtaining the lower natural mortality and the higher reproduction associated with a future large size. When the population is fished, the chance of cashing in on that investment becomes increasingly unlikely and it pays off to instead invest in reproduction here and now. The increasing investment in reproduction could have a large impact on asymptotic size. In chapter 3, we derived the relation between asymptotic size and reproductive investment (eq. 3.17)

$$W_\infty = \left(\frac{A}{k}\right)^4. \tag{6.13}$$

An increase in reproductive investment of 0.002 yr^{-1} (0.2 percent per year) will therefore lead to a reduction of asymptotic size by approximately $(1.002)^{-4}$— $1 \approx 1$ percent per year. This reduction is counteracted by the increasing investment in growth, however, although the increase is in most cases much smaller than the increase in reproductive investment. We might therefore expect fast reductions in asymptotic size. However, can we trust that estimation? chapter 3 also demonstrated an empirical relation between asymptotic size and size at maturation, with asymptotic size $W_\infty \approx w_m/0.28$ (fig. 3.4). Is asymptotic size determined

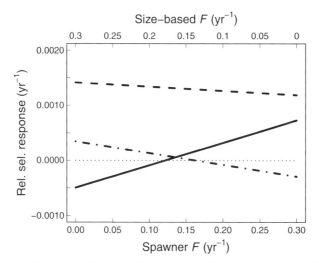

FIGURE 6.5. Selection response for a case III type selection (feeder versus spawner fishery). Pure size-selective fishing at the left edge (case I), and pure spawner fishery at the right edge (case III). $W_\infty = 20$ kg and trawl selectivity. Line types as in fig. 6.3.

by the size at maturation, or is it determined by the investment in reproduction? If asymptotic size is determined by size at maturation, then asymptotic size is expected to change only slightly, as the size of maturation changes much less than the investment in reproduction.

Whether asymptotic size is determined by maturation size or investment in reproduction did not really matter when we developed the trait-based demographic model in chapter 4 for the demographic impact assessment of fishing in chapter 5. There, the relations between asymptotic size and size at maturation and between asymptotic size and investment in reproduction could be used simultaneously. Now, with the quantitative genetics model, we need to know how w_m, A, and k conspire to determine asymptotic size. That we do not know. Observations of changes in asymptotic size are not much help either. It is quite difficult to estimate changes in asymptotic size in a fished population because fishing removes the large individuals, and therefore only few individuals actually reach asymptotic sizes. We can then turn to theory, but unfortunately it offers only limited advice. What we need is an explanation of why indeterminate growth, where $W_\infty > w_m$, emerges in fish instead of determinate growth, where $W_\infty = w_m$. Charnov et al. (2001) proposed that indeterminate growth emerges because it is possible to devote only a certain fraction of available energy to reproduction. The remainder will have to be used for growth, even though the individual is mature. This proposal is not very useful, as it does not offer a suggestion for what determines the upper limit to the fraction of energy devoted to reproduction. Another proposal by Thygesen et al. (2005) is tied to seasonal reproduction schedules by many fish. If fish are constrained to spawn only once a year, indeterminate growth with a fixed ratio between sizes at maturation and asymptotic sizes emerges. That theory also does not provide a solid suggestion for the amount of energy used for reproduction. Further, it still remains a conjecture, as no empirical follow-up has been made. The end result is that we do not know exactly how asymptotic size is determined by the three life-history traits: size at maturation, growth rate, and investment in reproduction.

6.4 SUMMARY: WHAT IS AN EVOLUTIONARY ENLIGHTENED FISHERIES MANAGEMENT?

In this chapter, we have developed a basic evolutionary impact assessment of fishing. The impact assessment estimated the selection responses resulting from size-selective fishing on three main life-history traits: size at maturation, growth rate, and investment in reproduction. The predicted selection responses from a fishing mortality comparable to F_{msy} are on the order of magnitude of 0.1 percent per year, smallest for size at maturation and largest for the investment in reproduction. The responses increase roughly proportional to the fishing mortality, so overfishing will not only result in depleted stocks and suboptimal yield production, but it will also lead to faster fisheries-induced evolution.

The estimated selection responses are about an order of magnitude smaller than empirical estimates, which hover around 1 percent per year (Jørgensen et al., 2007; Audzijonyte et al., 2013). The estimates are of similar order of magnitude than other theoretical estimates (Audzijonyte et al., 2013). The discrepancy means that either the empirical assessments overestimate the evolutionary changes or the theoretical estimates underestimate them. As discussed earlier, the empirical estimates are not rock-solid, as they are indirect estimates of genotypic changes based on observations of phenotypic changes, and are based on the assumption of a one-dimensional reaction norm. Eikeset et al. (2016) discusses how the failure to fully appreciate density-dependent effect led to overestimated rates in the North-east Arctic cod stocks. The theoretical models also have some weak spots. The predictions hinge on the values of the parameters and on the shape of the trade-offs. Differences in parameters are not enough to explain a factor of 10 underestimation, though (Andersen, 2010). For example, the heritability h^2 is fairly uncertain, but increasing it from 0.2 to the unrealistic value of 1 would only increase rates a factor of 5. The role of the trade-offs is less clear, and as discussed earlier, we do not have as much empirical support for them as we could like. The way to get at better understanding would be proper, controlled selection experiments that, unlike the

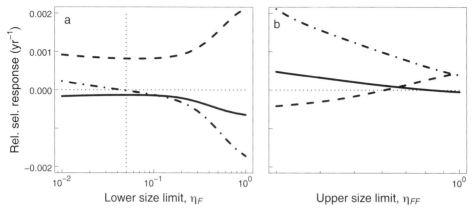

FIGURE 6.6. Selection responses for different size selectivity patterns: (a) trawl selectivity starting at size $\eta_F W_\infty$, and (b) trawl selectivity starting at $0.05 W_\infty$ and with an additional upper slot with zero fishing for sizes larger than $\eta_{FF} W_\infty$. The fishing mortality is adjusted such that the yield is the same for all selectivities. The yield is chosen arbitrarily as the one with a selectivity $\eta_F = 0.05$ and $F = 0.1$ yr^{-1}, corresponding to the vertical dotted line in panel a. Line types as in fig. 6.3.

experiments by Conover and Munch (2002), select for size and not directly on the trait, so we can compare theoretical and empirical estimates directly. Conover actually did that comparison, which is how he calculated the heritability. Such experiments are hard because the evolutionary rates would be much slower than in the case with direct selection. Until then, the estimates by models such as the preceding one, or more elaborate eco-evolutionary models (Eikeset et al., 2016), represent the best predictions we can make.

I have treated only size selective fishing. Increasing evidence is emerging that fishing may also select for behavior. Passive fishing gear, such as hooks, traps, or gill nets, might preferentially catch bold individuals that spend more time foraging than timid individuals (Arlinghaus et al., 2017). This type of selection on the fishery acts directly on the behavioral trait (boldness/timidity), and therefore it has the potential to impose a much faster selection response than the indirect selection imposed by a size selective fishery. The quantitative genetics model has been extended with a description of behavior (Andersen et al., 2018). Bolder individuals are associated with higher foraging rates (higher values of A) at the cost of a higher natural mortality and increased metabolic rates. That model indicates that selection against bold individuals indeed has the potential to reverse the selection responses on growth rates from being mostly positive to being negative. This means that passive fishing gear creates more timid fish with slower growth rates. Conversely, bolder individuals may be better at escaping active gear, such as trawl

(Diaz Pauli et al., 2015; Killen et al., 2015). Active gear may therefore select for bolder individuals with faster growth rates. As in the case of size selective fishing, the prediction of selection responses in a practical fishery, which often involves combinations of gear, will be complicated and has to be pursued on a case-by-case basis.

The evidence of evolutionary effects of fishing calls for an evolutionary enlightened fisheries management. The first step of this endeavor is to assess the evolutionary consequences of existing fishing practices. Such assessments can be performed quite simply by the procedure outlined in this chapter. The assessment requires knowledge of the life-history parameters of the species in question, which are typically fairly well known. Another requirement is knowledge of the size selection. This is often estimated as part of the standard stock assessment. The last requirement is knowledge of the mixture of feeder fishery and spawner fishery. This aspect of the fishing pattern is not part of standard stock assessments and would therefore require new knowledge. It should, however, be fairly straightforward to estimate this mixture simply as the ratio between catches landed inside and outside the spawning season. The assessment of selection responses can be improved by using more complicated eco-evolutionary models (Dunlop et al., 2009; Mollet et al., 2016).

A second aspect of an evolutionary enlightened management is to explore means to reduce the selection responses imposed by fishing. It has been proposed that a particular size selectivity will reduce the selection responses (Law, 2007), in particular one which does not fish the largest individuals (Jørgensen et al., 2009). In fig. 6.6, I explore how different selection patterns affect the selection respones. The fishing mortality is varied such that the fisheries yield is the same for all the patterns. Fishing only large individuals increases selection pressures (fig. 6.6a). Imposing a selection pattern with a slot that avoids catching the largest individuals (fig. 6.6 b) also seems to increase the selection responses. No matter the size selection pattern, the selection responses are of the same order of magnitude, around 0.1 percent per year. Therefore, the idea of reducing the effects of fisheries-induced evolution by not fishing the largest fish does not work. Taken together, it seems that there is no easy size-selection fix that will reduce the evolutionary impact of fishing (Matsumura et al., 2011). The only measure that clearly reduces selection responses is a lower fishing mortality.

CHAPTER SEVEN

Population Dynamics

The preceding three chapters developed a solid understanding of the demography of a fish stock and how it responds to fishing. The understanding was based on the assumption of a fish stock in steady state—that is, it neither grows nor decays in abundance. Fish, however, live in an unstable environment that affects all aspects of their life: larvae are subject to the vagaries of annual fluctuations, and adults face changes in food and predation. Most vulnerable are the early larval stages. In most cases, fish leave their offspring to fend for themselves right after they are spawned. The eggs and larvae are at the mercy of shifting currents that may carry them toward rich food sources, or may sweep them to deserts with little food. Of course, the adult population seeks out favorable spawning times and places, but they cannot predict the weather, only the average climatological conditions. A good example is the study of Baltic cod eggs and larvae by Hinrichsen et al. (2001). Simulating larval drift under different wind regimes, they showed how larvae in some years were transported toward their optimal habitat along the coast, while other years retained them in less optimal deep waters. In terms of the population model developed in chapter 4, such environmental variation in egg and larval survival makes the recruitment efficiency ε_R vary stochastically from year to year. This annual variability in recruitment means that the population is not always in steady state.

Besides the annual fluctuations in recruitment, a fish stock is also subject to changes in the abundance of food or predators. A change in the availability of food or predators may well persist over several years, so such changes often occur on slower time scales than the changes in recruitment. If such changes in the biotic environment are slow compared to the population dynamics of the fish stock, they may not be so important for the dynamics. In that case, the calculation of the dynamics of the population can be done under the assumption that the stock adjusts to the changes continuously and therefore essentially is in a steady-state situation. But how fast does a given fish stock respond to changes— which changes are "fast" compared to the internal dynamics, and which changes are "slow"? One measure of the rate of change of a fish stock is given by the reciprocal of the age of maturation: the longer it takes for a fish to mature, the slower the dynamics. The age at maturation scales metabolically, so the rate of change should be $\propto W_\infty^{-0.25}$ (see box 3.3). In other words: small species have a fast response, while large species are slow responders. However, we already saw in chapter 5 that we should be careful with relying on metabolic scaling rules when it comes to fish stocks. There, our expectation of the resilience of fish stocks based on metabolic scaling rules turned out to be wrong, and large species were about as resilient to fishing as small species. We therefore need to verify whether metabolic scaling rules are good proxies for the rates of response of fish populations or not.

Understanding the dynamics of fish stocks has important practical applications for fish stock management. Fisheries management is no stranger to collapsed fish stocks. It may even be argued that fisheries management has been shaped by the effort to recover collapsed fish stocks (van Gemert and Andersen, 2018b). A collapse mandates that a suitable recovery plan be drawn up. The plan typically involves a closure of the fishery or at least a reduction in fishing mortality. But how long do we need to close the fishery before we can expect to resume fishing? This question, again, involves a dynamic calculation.

The answers to these questions all involve different aspect of the dynamics of fish stocks—that is, a description of how their abundance and structure change over time. In mathematical terms, in chapters 4 through 6, we ignored the first term in the McKendrick–von Foerster equation (eq. 4.2), $\partial N / \partial t$, because we argued that the stock was in steady state and described as $N(w)$. I now put the time derivative back into the McKendrick–von Foerster equation and develop solutions that are functions of time: $N(w, t)$. As a first approximation, the speed of a stocks' response to changes can be approximated by the population growth rate in the absence of density dependence, $r_{\max}$. A population recovering from a depleted state experiences little density dependence so it will, at least initially, grow exponentially with the rate $r_{\max}$. Getting density dependence into the calculations requires numerical simulations of the full McKendrick–von Foerster equation, including a stock-recruitment function, but where, in contrast to chapter 4, the population is not in steady state but is allowed to change over time. I will show how the full dynamic response of a population can be deconstructed into three phases: an initial lag, the exponential growth phase, and the relaxation toward the equilibrium.

The game plan is as follows: I will first derive the population growth rate with various analytic and numeric approximations. Next, I will develop a full numerical solution to the McKendrick–von Foerster equations and use it to develop stylized recovery plans. Last, I will describe how a fish stock responds to fluctuations in the recruitment.

7.1   WHAT IS THE POPULATION GROWTH RATE?

The simplest possible dynamics population model is $dy(t)/dt = r_{max}y(t)$, where $y(t)$ is some measure of the population size, such as adult abundance or biomass, and r_{max} is the population growth rate with dimensions 1/time. The solution is exponential growth: $y(t) = y(0)e^{r_{max}t}$. In this solution, r_{max} is the exponent that determines the rate of increase—or decrease if $r_{max} < 0$. When we also want to resolve the population structure, we must find a solution in the form $N(w,t)$ that can be obtained from solving the time-dependent McKendrick–von Foerster equation (eq. 4.1)

$$\frac{\partial N(w,t)}{\partial t} + \frac{\partial g(w)N(w,t)}{\partial w} = -\mu(w)N(w,t). \tag{7.1}$$

BOX 7.1
ANALYTICAL APPROXIMATIONS OF POPULATION GROWTH RATE

Consider a population where juveniles invest all available energy into growth and where adults invest all available energy into reproduction. This assumption ignores that fish continue to grow after maturation, and to compensate partly for that I consider maturation at W_∞ and not at $\eta_m W_\infty$. With juvenile growth rate $g(w) = Aw^n$ (eq. 3.13) and mortality aAw^{n-1} (eq. 4.7), we obtain the solution to eq. 7.11 as

$$v(w) = Cw^{-n-a} \exp\left[-\frac{r_{max}}{A(1-n)}w^{1-n}\right], \tag{7.2}$$

where C is an integration constant. Notice that the solution is similar to the steady-state solution found earlier $N(w) \propto w^{-n-a}$ (eq. 4.21), just corrected with an exponential term that depends on the population growth rate. Applying the boundary condition of the McKendrick–von Foerster equation $R(t) = g(w_0)N(w_0,t)$ (eq. 4.22), and assuming that $R(t)$ increases exponentially with time as $R(t) = \tilde{R}e^{r_{max}t}$ gives:

$$v(w) = \tilde{R}\frac{w_0^a}{A}w^{-n-a} \exp\left[\frac{r_{max}}{A(1-n)}\left(w_0^{1-n} - w^{1-n}\right)\right]. \tag{7.3}$$

The solution, however, does not provide the population growth rate r_{max}. Getting r_{max} requires knowledge about the reproductive output of the adults. I will obtain the reproductive output with two approximations, as follows:

(*continued*)

(Box 7.1 *continued*)

1. The reproductive output $R(t)$ is simply the biomass flux of juveniles becoming adults, $g(W_\infty)N(W_\infty,t)W_\infty$, divided by the mass of an offspring, w_0, and discounted by the reproductive efficency ε_{egg} (eq. 3.19) and the recruitment efficiency ε_R (eq. 4.35):

$$R(t) = \frac{\varepsilon_{egg}\varepsilon_R}{w_0} g(W_\infty)N(W_\infty,t)W_\infty = \frac{\varepsilon_{egg}\varepsilon_R}{w_0} AW_\infty^n v(W_\infty)e^{r_{max}t}W_\infty. \tag{7.4}$$

Inserting $R(t)$ into eq. 7.3 and isolating r_{max} gives:

$$r_{max} = A\frac{1-n}{W_\infty^{1-n} - w_0^{1-n}} \left[(1-a)\ln(W_\infty/w_0) + \ln(\varepsilon_{egg}\varepsilon_R)\right]. \tag{7.5}$$

2. The preceding approximation ignored that adults may live to spawn several years. We can account for the adult life span by writing an equation for adult abundance M

$$\frac{dM}{dt} = g(W_\infty)N(W_\infty,t) - \mu(W_\infty)M, \tag{7.6}$$

where the first term on the right-hand side is the flux of juveniles becoming adults and the second term is loss to mortality. Making the *ansatz* $M(t) = M_0e^{r_{max}t}$ gives

$$M_0 = \frac{g(W_\infty)}{r_{max} + \mu(W_\infty)} v(W_\infty). \tag{7.7}$$

An adult uses all its available energy $g(W_\infty) = AW_\infty^n$ for reproduction, so the total reproductive output becomes

$$R(t) = \varepsilon_{egg}\varepsilon_R g(W_\infty)M(t)W_\infty/w_0 \tag{7.8}$$

$$= \varepsilon_{egg}\varepsilon_R v(W_\infty)\frac{A^2 W_\infty^{2n}}{r_{max} + aAW_\infty^{n-1}}\frac{W_\infty}{w_0}e^{r_{max}t}. \tag{7.9}$$

As before, the equation for $R(t)$ can be combined with the boundary condition to solve for r_{max}. Unfortunately, the solution is rather complex and given in terms of a Lambert-W function. Therefore, I will not write it here, but it can be obtained with standard symbolic mathematics software.

Notice how I consider growth $g(w)$ and mortality $\mu(w)$ to be constant over time; nonconstant growth and mortality are treated in chapter 10. We can solve the time-dependent McKendrick–von Foerster equation by assuming that the solution can be separated into independent functions of weight and time of the form

$$N(w,t) = v(w)e^{r_{max}t}. \tag{7.10}$$

It is tempting to assume that the dependency on weight, $v(w)$, is the steady-state solution from chapter 4 (box 4.3), and insert it in the *ansatz* (eq. 7.10). That is, however, not quite right. A growing population ($r_{max} > 0$) will have more juveniles per adult than a population in steady state, because these abundant juveniles are the ones that have just been spawned. The faster the population grows, the higher the imbalance in the ratio between juveniles and adults. In the size spectrum, this imbalance will be manifested as a number spectrum that declines faster with size than the steady-state spectrum. So we cannot use the steady-state solution $N(w)$ as a help to find the time-dependent solution $N(w, t)$, and we need to find a new solution for $v(w)$. Inserting the *ansatz* eq. 7.10 into eq. 7.1, we obtain the ordinary differential equation

$$r_{max} v(w) + \frac{dg(w)v(w)}{dw} = -\mu(w)v(w). \tag{7.11}$$

In box 7.1, I develop analytical solutions to eq. 7.11. The size structure of the population is approximated by (eq. 7.2)

$$v(w) \propto w^{-n-a} \exp\left[-\frac{r_{max}}{A(1-n)}w^{1-n}\right]. \tag{7.12}$$

The first term is the same as the steady-state solution for the juvenile spectrum $N(w) \propto w^{-n-a}$ found in box 4.3 (dotted line in fig. 7.1b). The exponential term is a correction that makes the spectrum decline faster with size than the steady-state solution, as anticipated. How much faster the spectrum declines depends on the ratio between the population growth rate r_{max} and the somatic growth rate A.

Box 7.1 also develops two approximations for r_{max}. In the simplest approximation, I assume that a populations' reproductive output is given by the flux of juveniles becoming adults. This approximation provides a simple approximation of the growth rate

$$r_{max} \approx A(1-n)W_\infty^{n-1}\left[(1-a)\ln(W_\infty/w_0) + \ln(\varepsilon_{egg}\varepsilon_R)\right], \tag{7.13}$$

where I have used the approximation that $W_\infty \gg w_0$ to simplify the solution from eq. 7.5.

The solution in eq. 7.13 reveals how the population growth rate depends on the traits and life-history parameters of a fish species: (1) the population growth is, not surprisingly, proportional to the growth rate of individuals A. (2) The term W_∞^{n-1} indicates a metabolic scaling of the growth rate leading to declining population growth rate with asymptotic size. (3) The first term in the brackets decreases if the physiological mortality a increases—increasing mortality leads to slower population growth. (4) The first term in the brackets also increases with asymptotic size and provides a logarithmic correction to the decrease in population growth

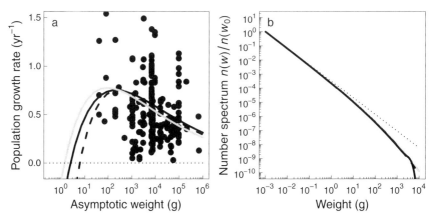

FIGURE 7.1. Population growth rates r_{max} and size spectra $\nu(w)$ of exponentially growing populations. (a) Population growth rates calculated by analytical approximation 1 (dashed line, eq. 7.13, almost hidden behind the solid line), analytical approximation 2 (gray line; see box 7.1), and the numerical solution (solid line). Data points from Hutchings et al. (2012), not corrected for temperature or growth rate A because of lack of information. (b) Number spectra for a species with $W_\infty = 10$ kg. The dotted line is $\propto w^{-n-a}$.

rate stipulated by the metabolic scaling term in front. (5) The second term in the brackets is negative, as ε_{egg} and ε_R are both less than one. Therefore, the lower the reproductive and recruitment efficiencies, the slower the population growth rate. (6) If the first term in the brackets is smaller than the second term, the population growth rate will be negative, and the population will go extinct. Extinction will happen if the physiological mortality is high, the asymptotic size is small, or the efficiencies are low. Taken together, the effect of the competing terms related to asymptotic size, the metabolic scaling that decreases with W_∞ and the logarithm of W_∞, is to produce a unimodal variation of population growth rate with asymptotic size, as shown in fig. 7.1a.

The time-dependent solutions in eq. 7.13 and eq. 7.12 are approximations where the adult life has been ignored. Finding a solution based on the detailed biphasic growth model developed in chapter 3 (eq. 3.18) requires numerical simulations. That involves turning the continuous McKendrick–von Foerster equation into a discrete matrix equation of the form

$$N^{t+1} = \mathbf{A}N^t, \tag{7.14}$$

where the superscripts refer to the time steps. How to discretize the continuous McKendrick–von Foerster equation to determine the matrix $\mathbf{A}$ is described in box 7.2. The matrix elements are determined by the a combination of growth rate and mortality. I use the same formulation as for the steady-state demography

BOX 7.2

POPULATION TRANSITION MATRIX

Solving the dynamic McKendrick–von Foerster equation requires a numerical scheme that can handle the time dependency of the solution. I use a standard finite-difference solution scheme that is commonly used to solve hyperbolic partial differential equations.

The solution is discretized on a logarithmic grid starting with the first grid point w_1 and the following grid points as $w_j = (1 + c_{exp})w_{j-1}$. The factor c_{exp} determines the expansion of the grid. I have used $c_{exp} = 0.1$ for the simulations presented here, which gives about 200 grid points. The time derivative is discretized with a central difference scheme and the derivative with weight uses an upwind scheme

$$\frac{N_j^{t+1} - N_j^t}{\Delta t} + \frac{g_j^t N_j^{t+1} - g_{j-1}^t N_{j-1}^{t+1}}{\Delta w_j} = -\mu_j^t N_j^{t+1}, \qquad (7.15)$$

with $\Delta w_j = w_j - w_{j-1}$ and $\Delta t \approx 0.1$ yr being the time step. Subscripts are grid numbers and superscripts are time steps. Collecting terms of N_j and N_{j-1} gives:

$$N_{j-1}^{t+1} \underbrace{\left(-\frac{\Delta t}{\Delta w_j} g_{j-1}^t\right)}_{A_j} + N_j^{t+1} \underbrace{\left(1 + \frac{\Delta t}{\Delta w_j} g_j^t + \Delta t \mu_j^t\right)}_{B_j} = N_j^t \qquad (7.16)$$

where I have defined the coefficients evaluated at the previous time step t as A_j and B_j. The reproductive output of weight group j is simply $R_j = \varepsilon_R \psi_m(w_j) R_{egg}(w_j) \Delta w_j / w_0$, where R_{egg} is the reproductive output per time by fish of size w_j (see eq. 4.35). We can now define the population transition matrix $\mathbf{A}$ by putting the coefficients B in the diagonal, A on the subdiagonal, and R_j in the first row.

in chapter 4: growth rate and reproductive output from the biphasic growth model (eq. 3.18) and (eq. 3.19), and declining mortality with size (eq. 4.7).

Once the McKendrick–von Foerster equation is brought into matrix form, we can use standard techniques from linear algebra to find the population growth rate r_{max}. First, the solution to eq. 7.14 is written as an eigenvalue decomposition of $\mathbf{A}$

$$N^t = \sum_i \lambda_i^t v_i, \qquad (7.17)$$

with as many eigenvalues λ_i and eigenvectors v_i as there are elements in the vector N^t. In the long run, the sum will be dominated by the eigenvector associated with the eigenvalue with the largest real value, λ_*

$$\lim_{t \to \infty} N^t = \lambda_*^t v_*. \tag{7.18}$$

Here, λ_* is raised to the power t, so t is not an index as it is on N. Notice the similarity with the *ansatz* (eq. 7.10) that was used to solve the McKendrick–von Foerster equation: v_* plays the role of $v(w)$, and the relation between λ_* and r_{max} is found by solving

$$\lambda_*^t = e^{r_{max}t} \Leftrightarrow r_{max} = \ln(\lambda_*). \tag{7.19}$$

Fig. 7.1a shows how the full numerical solution compares with the simple approximation from eq. 7.13. The simple approximation works surprisingly well: species with asymptotic sizes below about 1 gram have negative population growth rates, species with an intermediate size in the range 10 to 100 grams have the highest population growth rates, and for larger species the population growth rates declines slightly with asymptotic size. Consequently, the medium-size species are the least likely to be threatened, in accordance with observations (Ripple et al., 2017). The magnitude of the growth rate is on the order of 0.5 yr^{-1}. That fits fairly well with estimations of growth rates made by Hutchings et al. (2012), shown as points in fig. 7.1a. Some of the large variation in the observed population growth rates is because of differences in somatic growth rate A between the species. I have been unable to calculate A for the species in the data set, so I could not correct for this. Population growth rates of this magnitude double the population size in about a year (doubling time is $\ln(2)/r \approx 1.4$ yr^{-1}). This is a remarkably fast rate of increase for large vertebrates, for which metabolic arguments would predict rather slow growth.

7.2   HOW FAST DOES A POPULATION RECOVER FROM OVERFISHING?

While the population growth rate is a good first approximation of stock dynamics, a full dynamic solution is likely to differ in several ways. As an example, let us consider a recovery scenario where a depleted stock is allowed to increase in abundance. During the recovery, the stock structure $v(w)$ will change, and will not be invariant as assumed in the population growth rate analysis. Further, at some point density-dependent effects will kick in and limit population growth. Including these effects into the population dynamics requires a full numerical solution of the McKendrick–von Foerster equation. The numerical scheme for a dynamic model is given in box 7.3.

The dynamic scenarios in fig. 7.2 on a small and a large species illustrate how population growth can be divided into three phases shown with numbers on a gray

---

BOX 7.3

TIME-DEPENDENT NUMERICAL SOLUTION

The solution of the time-dependent McKendrick–von Foerster equation follows directly from the discretization in box 7.2. We could apply eq. 7.14 recursively to iterate the solution as $N^{t+1} = \mathbf{A}^{-1} N^t$, but that is slow. Here, I develop a faster algorithm that is also suitable for use when growth, reproduction, and mortality vary with time. This will be useful in part IV. We can rewrite eq. 7.16 as a recursive relation for N_j^{t+1}

$$N_j^{t+1} = \frac{N_j^t - A_j N_{j-1}^{t+1}}{B_j}. \tag{7.20}$$

The recursion can be solved with an iterative procedure once the value of the spectrum at the first grid cell, N_1^{t+1}, is known. That value can be found from the boundary condition

$$N_1^{t+1} = \left(N_1^t + \frac{\Delta t}{\Delta w_1} R_{\text{tot}}\right) \frac{1}{B_1}, \tag{7.21}$$

where $R_{\text{tot}} = \sum_j R_j$ is the total reproductive output.

This scheme is technically known as a semi-implicit first-order upward scheme— see the excellent chapter on numerical solutions of partial differential equations in the classic *Numerical Recipes* by Press (2007). The scheme is simple, and it does the job fairly well. It does have a flaw, however: it produces numerical diffusion. This diffusion does not matter much if the solution is smooth, as it is in all the cases presented here, but if the solution contains sharp gradients or shocks, the scheme will fail to deliver an accurate solution. Such situations may occur if the solution develops cohort cycles where the population is dominated by a single cohort of individuals spawned at the same time (Persson et al., 1998). In such cases, one needs to use a higher order scheme with a limiter (see Zijlema, 1996) or drop finite difference methods entirely and use a cohort-oriented solution—technically known as "characteristics" in the general literature on numerical solutions of hyperbolic partial differential equations, or as the "escalator boxcar" when applied to size-structured populations (De Roos, 1988).

---

background. First, there is a delay before the new recruits become mature. The delay is the age of maturation, which scales metabolically with asymptotic size $\propto W_\infty^{1-n}$ (eq. 3.25). After the initial recruits have matured comes a phase of exponential growth, evident by the straight line on the semi-log axis. The slopes in the exponential phase of the two species with different asymptotic sizes are similar, which is also to be expected from the calculation of the population growth rates.

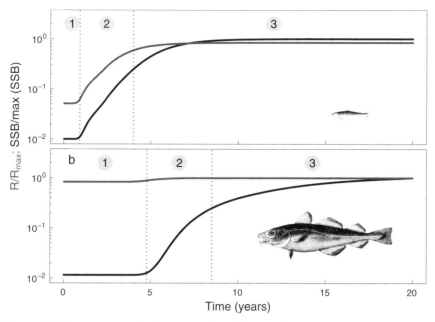

FIGURE 7.2. Recovery scenarios for two species with $W_\infty = 20$ g (top) and $W_\infty = 20$ kg (bottom), here exemplified as an anchovy and a cod. Spawning stock biomass is shown in black, and recruitment in gray. Three phases are indicated with numbers and vertical dashed lines: (1) the initial lag until newly hatched larvae mature, (2) exponential growth, and (3) slow relaxation toward full recovery.

When the exponential phase ends, population growth slows down, while the population structure adjusts itself to the final steady-state population structure. This last phase is substantially longer for the large species than for the small species, because the duration is again determined by the age at maturation, so it scales metabolically with asymptotic size.

A more practically relevant scenario is the recovery of an overfished population. As an example, I consider a species with $W_\infty = 20$ kg in fig. 7.3. The stock is unfished until it is discovered and immediately fished with a trawl selectivity and a fishing mortality $F = 1.5$ yr^{-1} over 15 years. During this period, the spawning stock biomass is reduced to almost nothing. Initially, the yield is tremendously high, much higher than what could be achieved by MSY exploitation—this is the boom of a newly discovered stock. After a few years of bountiful fishing, the yield declines rapidly, but it is still at about half the maximum sustainable yield. Some time after the stock has almost disappeared, a recovery plan is put into place, and a moratorium is introduced. Immediately, the yield drops and the spawning stock biomass begins to recover. After six years, it reaches the level that corresponds

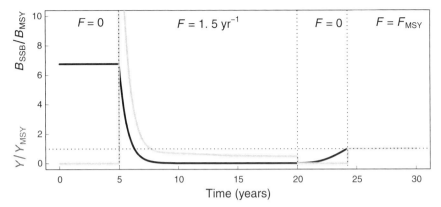

FIGURE 7.3. Illustration of the development of a fishery on a stock with $W_\infty = 20$ kg, showing four phases: unexploited, overexploited, moratorium, and fishing for maximum sustainable yield. Fishing is with a trawl selectivity, as shown in fig. 5.3.

to the steady-state biomass under maximum sustainable yield exploitation. Notice how the initial delay seen in fig. 7.2 is virtually absent here. This is because only the largest individuals are fished, so the initial delay is the time it takes to grow from the size at 50 percent fishing to maturing—from W_F to W_∞ which is short. Notice also how fishing at the maximum sustainable yield entails a substantial reduction of the spawning stock biomass—fishing for the maximum sustainable yield does have a significant impact on the stock.

The scenario in fig. 7.3 is stylized but not unreasonable. Both Baltic and North Sea cod were exploited with rates above 1 yr^{-1} in the 1980s and 1990s, until the stocks collapsed. After struggling to maintain a recovery plan, there are now signs that at least the North Sea cod is recovering. Establishing a moratorium in a recovery plan is difficult in practice, as there will be a pressure to allow some fishing. The absence of an income over a period of six years will have disastrous consequences for the industry, which will most likely shut down or leave to develop new fisheries. Continuing some fishing will allow the remaining fishing fleet to stay operational. Therefore, moratoria have a hard time gaining support from the fishing industry, who will be unwilling to support a measure that puts them out of business even if they see the possible long-term benefit. The fishers have probably become accustomed to the low yields from the overfished stock, and while they may pine for the high yields of earlier times, they will probably be too economically stressed to forgo yield entirely for a longer periods. Also, fishers typically have a high internal discount rate, so they would rather keep the current low yields now than stopping fishing on a biologists' promise of future higher yields. Further, reaching MSY exploitation will not return the golden age of the

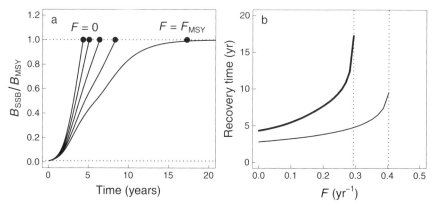

FIGURE 7.4. Recovery of overfished species to MSY levels. (a) Spawning stock biomass of a stock with $W_\infty = 20$ kg starting from a state where it is fished with $F = 1.5$ yr^{-1}, and then fished during the recovery phase with mortalities ranging from 0 to F_{msy}. (b) Time to recovery to B_{msy} as a function of the fishing mortality during the recovery period, for species with $W_\infty = 20$ g (thin line) and $W_\infty = 20$ kg (thick lines). The vertical lines show the F_{msy} of each stock.

initial overfishing—in the example, it promises only a factor of 2 higher yields. Last, even if no direct fishing would be allowed during the recovery period, it is likely that the population will suffer some fishing mortality due to illegal fishing and by-catch in other fisheries. Therefore, a recovery plan that does not consider some fishing mortality during the recovery is in most cases unrealistic.

The problem is of course that fishing during the recovery delays the recovery. Fig. 7.4b shows the length of the recovery period for two species ($W_\infty = 20$ g and 20 kg) as a function of the fishing mortality during the recovery period. Clearly, fishing slows down the recovery—in particular, when the mortality is close to F_{msy}. What is also interesting is that the recovery period of the large species is only about a factor of 2 longer than the recovery of the small species. This difference is much less than anticipated by metabolic scaling rules, which should give a factor of $(20,000/20)^{0.25} \approx 5.6$ in difference between the two species. A more complete analysis would include the economics of the fishing industry and account for the profit during the recovery and include the potential higher future yield, weighted by the discount rates. Such an analysis has been conducted by Calduch-Verdiell et al. (2011).

The recovery scenarios developed here provide an optimistic view on fish stock recovery. The biggest oversight is ignoring Allee effects, also referred to as *depensation* or *inverse density dependence*, whereby population growth rates are depressed when the stock size dips below a certain threshold. There are good reasons to give such Allee effects serious consideration, as many fish stocks have indeed failed to recover as expected. Allee effects can occur when other species invade while the stock is depleted and partly take over the stock's ecological niche. Another possibility is an increase in natural mortality originating from natural predators that consume an increasingly large fraction of the stock as it is decimated by overfishing. This effect has been suggested as one reason for the failed recovery of the Northern cod off Labrador and Newfoundland, where seals feed on cod (Benoît et al., 2011). Last, Allee effects may emerge due to the combination of competition and predation between a large piscivore and its forage fish prey, as has been suggested for the Baltic Sea cod (van Leeuwen et al., 2008). Such ecosystem effects are hard to detect while they are occurring, and even harder to predict in advance. It is therefore difficult to include them explicitly in a recovery plan. The possibility of an Allee effect should, however, be acknowledged, and model predictions such as those in fig. 7.4 must be considered optimistic estimates that provide only a lower bound on the time it takes for a population to recover.

7.3   HOW DOES A POPULATION RESPOND TO ENVIRONMENTAL FLUCTUATIONS?

We have seen how a population responds to a press perturbation where the fishing mortality is rapidly changed and then maintained at the new level until the population has reached a new steady state. Now I will consider the dynamic response to a perturbation that is varying continuously. The ecologically relevant example is how the annual variability in the recruitment affects the adult population. Here comes one of the cases where the continuous-time formulation makes things more complicated than in the classic age-based formulation. In the age-based formulation it would be simple to multiply the annual recruitment with a random number. When time is continuous, we need to create a continuous-time random process with similar properties to the annual random recruitment.

To simulate the annual variability in the success of spawning, I add noise to the recruitment efficiency. The noise is "red," which means that it autocorrelated in time. I use a correlation time of 1 year to mimic how conditions affecting recruitment changes from one year to the next. Such an autocorrelated noise $x(t)$ can be modeled as an Ornstein-Uhlenbeck process with the stochastic differential equation

$$dx(t) = -\tau x(t)dt + \sigma_x dW(t). \tag{7.22}$$

Here, $\tau = 1$ year is the autocorrelation, $\sigma_x = 2$ is the spread of the process, and $W(t)$ is a Brownian motion process (also called a Weiner process). Why does the equation have this strange form—why don't I just divide with dt on both sides to write it in the standard form of a differential equation? This is because the Brownian motion process varies on infinitesimally small time intervals, so its derivative is

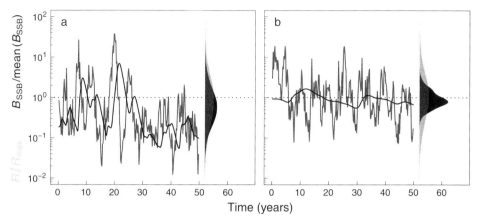

FIGURE 7.5. The influence of noise on recruitment (gray) and on spawning stock biomass (black), shown for two species with asymptotic size $W_\infty = 20$ g (a) and $W_\infty = 20$ kg (b). The gray and black regions on the right side of each panel show the probability density functions of the noise on recruitment and spawning stock biomass.

undefined. To acknowledge this issue, it is standard practice to write stochastic differential equations in this particular way. We need to solve the Ornstein-Uhlenbeck process in discrete time, where it becomes

$$x(t + \Delta t) = x(t) - \tau x(t) \Delta t + \sigma_x \sqrt{\Delta t} \mathcal{N}, \qquad (7.23)$$

where $\mathcal{N}$ is a normal distributed random number with 0 mean and 1 variance. I then model recruitment efficiency as

$$\varepsilon_R(t) = \varepsilon_R e^{x(t)}. \qquad (7.24)$$

This procedure leads to a recruitment that varies roughly a factor of 10 between years, as seen with the gray lines in fig. 7.5.

The adult age structure acts as an averaging operator on the noisy recruitment. In a large-bodied species the adults are a mix of many cohorts, each deviating from the expected steady state solution according to the noise on the recruitment on the cohort. When integrating over the adult size range to find the spawning stock biomass, some of the variation between the cohorts will be averaged out by the many cohorts. The degree of averaging will be determined by the number of cohorts in the adult population: small species will have few cohorts and little averaging, while large species will have many. This effect is evident in fig. 7.5: the spawning stock biomass of the small species almost follows the noisy recruitment, just with a delay corresponding to the age at maturation (here, 1.4 years), while the large species varies much less. This averaging by the adult cohorts is one reason why the biomass and catch of small species typically vary much from year to year. The other reason, as discussed in section 4.2, is that small species are more susceptible to recruitment variability than larger species due their lower degree of density dependent regulation.

7.4 SUMMARY

The dynamic response of a fish population to perturbations is complex. The response will depend on the exact circumstances of the perturbation, such as the size range that is affected and the temporal nature of the perturbation. Nevertheless, this chapter has developed a general understanding of how a fish stock responds to idealized perturbations: a press perturbation and a continuously varying perturbation. From these examples, we can make general expectations of the responses to perturbations in general. Three rules emerged from the examples: lag, averaging, and nonmetabolic scaling of population growth rate.

First, there is a lag in the population dynamics associated with the time to maturation. The age at maturation scales metabolically $\propto W_\infty^{1-n}$, so the length of the lag increases with the asymptotic size of the species. The lag may be important in an initial response to a perturbation, as shown in fig. 7.2. The lag also determines the slow relaxation to the final population structure that happens once a population has recovered to the point where it is settling into the final steady state population structure.

Second, rapid perturbations will be averaged out by the adult cohort. In this case, *rapid* means fluctuations on a time scale shorter than the number of dominant cohorts in the adult population. The number of dominant cohorts will again scale metabolically; species with larger asymptotic size will have more adult cohorts than small species. Consequently, larger species will be better at smoothing annual fluctuations in recruitment than smaller species.

The two rules about lag and averaging could be anticipated a priori, at least in general terms, but the theory quantifies their effects. The third rule is more counterintuitive, because it states that the population growth rate does not obey metabolic scaling rules. Rather, population growth rates are roughly independent of asymptotic size, as seen in fig. 7.1a. We have seen this kind of nonmetabolic behavior before, in section 5.3 for the fisheries reference points (fig. 5.8). The population growth rate is the most important quantity that describes how fast a population is able to respond. The high growth rates of particularly the large species is testament to the remarkable resilience of fish populations. Why do metabolic scales rules not work as anticipated for fish? I will return to this question in the next chapter, where I compare elasmobranchs and teleosts. Spoiler alert: elasmobranchs actually *do* obey metabolic scaling rules. Anyway, it is not key here to know why (teleost) fish do not obey metabolic scaling rules, the main thing is to remember that metabolic scaling rules are not always to be trusted for fish.

The population growth rate as given in eq. 7.13 has more uses than just being a fancy analytic approximate solution of the McKendrick–von Foerster equation. It can provide decent approximate values of the population growth rates, to be used in unstructured models such as the logistic growth model

$$\frac{dB_{SSB}}{dt} = r_{max} \left(1 - \frac{B_{SSB}}{B_{SSB.max}} \right) B_{SSB}, \tag{7.25}$$

where $B_{SSB.max}$ is the B_{SSB} in the absence of fishing. The approximation of r_{max} requires knowledge of the growth rate constant A, the physiological mortality a, and the asymptotic weight W_∞. These three parameters can be derived from standard von Bertalanffy growth parameters K and L_∞, and from the adult mortality M using the relations (eq. 3.8) for A, (eq. 4.42) for a, and table 2.1 for W_∞. The unstructured model is much easier to solve than the McKendrick–von Foerster equation, but it is unable to capture the two lag-phases of the population dynamics well because it does not resolve the changes in the actual stock structure explicitly.

PART III
TRAITS

CHAPTER EIGHT

Teleosts versus Elasmobranchs

Even though this book is about fish, I have yet to define exactly what I mean by a "fish." In the data analyses I have largely considered only teleosts (Teleostei), but the common definition of fish also includes hagfish, lampreys, cartilaginous fishes (Chondrichthyes), and other bony fish: lobe-finned fish (Sarcopterygii), Holostei, and Cladistia. Among them, teleosts represent by far the dominant group, in terms of both biomass and living number of species—approximately 26,000. Second in line comes the cartilaginous fishes, where elasmobranchs (sharks, rays, skates, and sawfish) dominate. In this chapter, I look into the differences and similarities between the two largest groups of fish: the teleosts and the elasmobranchs.

Teleosts dominate the marine size spectrum in the size range from about 1 g to 100 kg. Even though elasmobranchs are also present in the larger end of that body size range, they are not dominant: the global biomass of teleost fish is on the order of 9×10^8 tons, while elasmobranch biomass is a factor 100 times lower, at around 8×10^6 tons (Jennings et al., 2008; though some tropical systems have a higher relative mass of elasmobranchs). In other words: the average fish is a teleost.

From a production perspective, as a source of food or protein for other food production, we mostly think of teleosts. Elasmobranchs are not considered a food resource to the same degree. In the Western world, they are mostly a culinary curiosity, while in Southeast Asia they are a rare delicacy, such as shark fin soup. Due to their low abundance, only small fisheries are directed toward elasmobranchs and about 50 percent of the catches are taken as by-catch in fisheries for teleosts. Culturally, elasmobranchs are a source of wonder and awe. Watching Jacques Cousteau documentaries as a child, the fearsome sharks and outworldly elegant rays caught my attention much more than (teleost) fish schools. Not to mention *Jaws!*—it is hard to imagine a movie called *Herring!* filling the box office. Ecologically, teleosts and elasmobranchs play somewhat different roles. Sharks are top predators feeding typically on fish, though the largest sharks are more similar to baleen whales and feed on zooplankton. Oddly, the largest elasmobranchs thus compete with the smallest teleosts, the forage fish. While teleosts are very resilient to fishing, elasmobranchs are generally considered particularly sensitive top predators (Stevens et al., 2000). Rays and skates, for example, have been fished to almost extinction in large parts of the North and Irish Seas (Brander, 1981) and the Northwest Atlantic (Baum et al., 2003).

This chapter describes the differences between teleosts and elasmobranchs from a population dynamics perspective. As we will see, the main difference between the two groups is in their offspring size strategy: teleosts make small offspring; elasmobranchs make large offspring. I will use this difference to quantify the sensitivity of elasmobranchs to fishing relative to teleosts. Last, I develop an evolutionary explanation for why the offspring size strategy differs between teleosts and elasmobranchs.

8.1   HOW DO TELEOSTS AND ELASMOBRANCHS DIFFER?

Teleosts and elasmobranchs are remarkably diverse in size and shapes. Some are torpedo shaped and built for speed, such as many sharks and round-fish—in particular, scombroids. Others are flat, such as flounders, or skates and rays. Regarding size, there is a remarkable difference between teleosts and elasmobranchs: teleosts range from the tiny anglerfish (*Photocorynus spinicep*) with an asymptotic size around 1 g, to the giant oarfish (*Regalecus glesne*), measuring 7.6 m or more, and ocean sunfish (*Mola mola*) weighing over 1 tons. In contrast, the smallest shark is the dwarf lanternshark (*Etmopterus perryi*), with asymptotic size 100 g, and the largest elasmobranch is the whale shark (*Rhincodon typus*), weighing in at over 10 ton. Elasmobranchs occupy a smaller size range than teleosts, and there is a remarkable absence of small elasmobranchs; there are no "forage sharks."

Physiologically, there are fundamental differences between teleosts and elasmobranchs. Teleosts have a bony skeleton; elasmobranchs have a cartilaginous skeleton. Elasmobranchs have 5 to 7 gill openings, while teleosts have only 1. Elasmobranchs have no swim bladders, while many teleosts (though not all) do. The list of differences in physiology continues in increasing detail; however, what matters for the population dynamic theory developed in this book is not these detailed differences, but rather how the differences are manifested in the main traits and life-history parameters. Fig. 8.1 compares the four main life-history parameters: growth rate parameter A, ratio between size at maturation and asymptotic size η_m, specific reproductive output $\varepsilon_{egg}k$, and physiological mortality a. Specific reproductive output is expected to scale with $W_\infty^{-0.25}$, so I have multiplied by $W_\infty^{0.25}$ to create a measure of reproductive output invariant of asymptotic size (see eq. 3.17 and fig. 3.5). Despite the profound differences in physiology between teleosts and elasmobranchs, none of the four life-history parameters are significantly different.

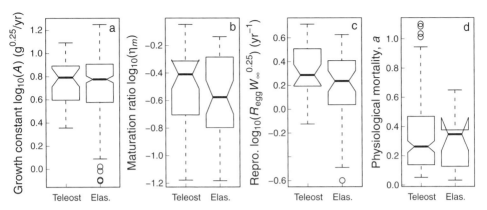

FIGURE 8.1. Comparison between life-history parameters of teleosts and elasmobranchs. The thick horizonal lines show the median, and the notches indicate whether the two groups are significantly different: if the notches overlap, the two medians are not significantly different. (a) Growth rate coefficient corrected to $15°C$ with a $Q_{10} = 1.83$ (see also fig. 3.3); (b) ratio between weight at maturation and asymptotic weight η_m (see also fig. 3.4); (c) reproductive output corrected with $W_\infty^{0.25}$ (see also fig. 3.5); (d) physiological mortality a calculated from the ratio between adult mortality and von Bertalanffy growth constant, M/K, using eq. 4.42 (see also fig. 4.6). Data are from Olsson and Gislason (2016) with extra data for A from Kooijman (2000), and teleost mortality data from Gislason et al. (2010).

The aspect where teleosts and elasmobranchs really differ is in the offspring size strategy (fig. 8.2). Teleosts make small offspring, around 1 mg, irrespective of their asymptotic size: anchovies make small eggs, cod make small eggs, even bluefin tuna make small eggs. In contrast, elasmobranchs make eggs proportional to the asymptotic size, roughly 370 times smaller than the asymptotic size or 100 times smaller than the size at maturation. There are also differences in whether offspring are in the form of eggs or live offspring, where elasmobranchs show a stronger preference for live offspring than teleosts. It is interesting to note that smallers sharks seems to prefer making eggs, while larger sharks make live offspring. One can speculate that the upper size of eggs is limited by the speed at which oxygen can diffusive towards an egg, which increases with the radius of the egg (Munk and Riley, 1952), while the oxygen requirements scales metabolically. However, whether offspring are in the form of eggs or live offspring does not influence the offspring size strategy.

The final life-history parameter is the recruitment efficiency, ε_R. This parameter cannot be inferred directly from physiology or growth curves. There are reasons to expect that the recruitment efficiency is larger in elasmobranchs than in teleosts. The recruitment efficiency represents the overall hatching success of eggs and initial survival of offspring. The parameter cannot be larger than 1. Elasmobranchs

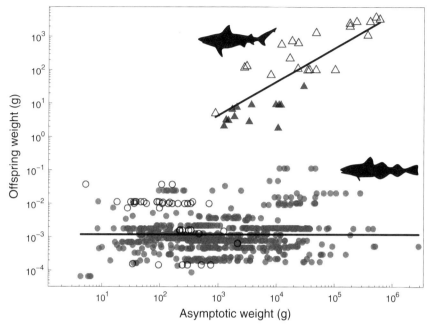

FIGURE 8.2. Offspring size for teleosts (circles) and elasmobranchs (triangles) as a function of asymptotic size. Oviparous (egglaying) species are shown with filled symbols and viviparous (live offspring) with open symbols. The geometric mean of teleost offspring size is 1.1 mg; the mean offspring:asymptotic size ratio for elasmobranchs is $w_0/W_\infty = 0.0044$. Data for teleosts from FishBase (Froese and Pauly, 2017); data for elasmobranchs from Olsson and Gislason (2016).

make relatively fewer offspring, so they are able to spawn them carefully in places that are conducive to their development and survival. In contrast, teleosts rely on external fertilization, so most species have to spawn their eggs freely in the water masses to have them fertilized. Though they select the most optimal time and place to spawn, the further fate of the eggs and larvae are subject to the current and temperature conditions in the given year. In addition, a higher proportion of elasmobranch species make live offspring than teleost species (fig. 8.2). This strategy eliminates the risk of egg mortality and will also account for a higher recruitment efficiency. I will use $\varepsilon_R = 0.3$ for elasmobranchs, and later, in fig. 8.3 and section 8.4, I provide a justification for this value.

In conclusion: no systematic differences can be distinguished between teleosts and elasmobranchs in growth, reproductive output, or mortality. The difference between the two groups are chiefly in the offspring-size strategy, with teleosts pursuing a many-small-eggs strategy, while elasmobranchs make offspring roughly a factor 1:100 smaller than size at maturation. Other differences are the range of

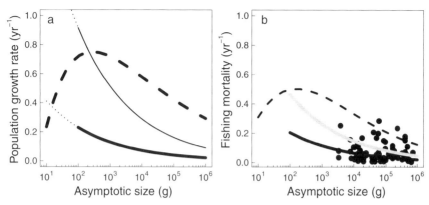

FIGURE 8.3. Population growth rates and fisheries reference points for elasmobranchs (solid lines) and teleosts (dashed lines). The offspring size of sharks is $w_0 = 0.0044W_\infty$. (a) Population growth rate from the simple approximation in eq. 7.13, and reproductive efficiency $\varepsilon_R = 0.3$ (thick line) and $\varepsilon_R = 1$ (thin line). The lines are dotted for W_∞ smaller than the smallest known sharks to illustrate the growth rates that smaller sharks could potentially achieve. (b) Fishereries reference points F_{msy} (black line) and F_{crash} (gray line), both calculated with $\varepsilon_R = 0.3$ for sharks and $\varepsilon_R = 0.03$ for teleosts. F_{crash} for fish is out of the scale and therefore not shown; see fig. 5.8. Data points from Zhou et al. (2012).

asymptotic sizes, where there are no elasmobranchs smaller than 100 g, and in the recruitment efficiency where we expect that elasmobranchs' offspring size strategy to be more efficient than the teleost strategy.

8.2 HOW SENSITIVE ARE ELASMOBRANCHS TO FISHING?

Elasmobranchs are generally thought to be more sensitive to fishing than teleosts (Holden, 1973). This is partly the reason why there is little directed fishery towards elasmobranchs. Another reason is their lower abundances, which leads to low yields. We can use the procedures developed in chapters 5 and 7 to assess the sensitivity of elasmobranchs to fishing. I will use the same life-history parameters for elasmobranchs as for fish (from table A.2, except for the offspring size, where $w_0 = 0.0044W_\infty$ and the reproductive efficiency $\varepsilon_R = 0.3$).

The population growth rate can be approximated by the simple relation derived in eq. 7.13

$$r_{\max} \approx A(1-n)W_\infty^{n-1}\left[(1-a)\ln(W_\infty/w_0) + \ln(\varepsilon_{\text{egg}}\varepsilon_R)\right]. \qquad (8.1)$$

For elasmobranchs, the ratio between asymptotic size and offspring size, W_∞/w_0, is constant (fig. 8.2). Because of this invariant, the term in the angular brackets is independent of W_∞, and the population growth rate scales according to metabolic scaling rules (box 4.1): $r_{max} \propto W_\infty^{n-1}$ (fig. 8.3a). Population growth rates decrease much faster with increasing asymptotic size than for teleosts. Using the same recruitment efficiency for elasmobranchs as for teleosts ($\varepsilon_R = 0.03$), leads to a negative population growth rate. This result is the first solid indication that the recruitment efficiency indeed has to be higher for elasmobranchs than for teleosts. The upper limit for the recruitment efficiency is $\varepsilon_R = 1$, and even in that case teleosts have a higher population growth rate than elasmobranchs. However, elasmobranchs with small asymptotic size seem to have similar or even higher population growth rates than teleosts. Why, then, are there no smaller elasmobranchs? I will return to this question later, but first I will look at the effect of fishing elasmobranchs.

I have exposed elasmobranchs to fishing with the same trawl size-selectivity used in chapter 5 on teleosts. I have then calculated two fisheries reference points: the fishing mortality that gives the highest yield, F_{msy}, and the fishing mortality where the population crashes, F_{crash} (fig. 8.3b). As with the population growth rate, it is evident that elasmobranchs on average tolerate much lower fishing pressures than teleosts. No wonder that even by-catch rates are sufficient to drive elasmobranchs to extinction in systems with high fishing pressure. It should be noted that the rates of population growth and fishing mortality are for a temperature of 15°C, as per the temperature correction of the growth coefficient A. As most elasmobranchs live in warm waters, they will also be able to tolerate higher fishing pressures because of the expected faster growth rates due to higher temperatures.

Elasmobranchs clearly lose the battle for resilience to teleosts. Why does a difference in offspring size have such a strong effect that it cannot be compensated by a 10 times higher recruitment efficiency? We can explore this question by looking at how the cohort biomass increases in a population. The cohort biomass per offspring at size w can be approximated as the probability to reach that size multiplied by the size itself: $B_{cohort} = P_{w_0 \to w} w$. Survival is calculated as (eq. 4.31)

$$P_{w_0 \to w} = \exp\left[-\int_{w_0}^{w} \frac{\mu(\omega)}{g(\omega)} \, d\omega \right] = \left(\frac{w}{w_0} \right)^{-a}, \tag{8.2}$$

where I have used juvenile growth Aw^n and mortality aAw^{n-1} (eq. 4.7). Multiplying by the size w we get that the cohort biomass per offspring scales as $(w/w_0)^{1-a} \approx (w/w_0)^{1-0.42}$. A shark, irrespective of its asymptotic size, will increase its cohort biomass by a factor $(100)^{0.58} \approx 14$ when it reaches maturation, while a teleost with asymptotic size 100 kg will increase its cohort biomass by a factor of $(100 \, kg/1 \, mg)^{0.58} \approx 20,000$. Clearly, even the factor of 10 higher recruitment efficiency of sharks is unable to compensate for that difference in cohort biomass.

8.3 WHY DO TELEOSTS MAKE SMALL EGGS?

It is now evident that the teleosts small-egg strategy is superior to the strategy of making large offspring adopted by elasmobranchs. Teleosts make small eggs because it makes evolutionary sense—making small eggs is the offspring size strategy that optimizes an individuals' fitness. But why it is optimal to follow this strategy? After all, it is not so obvious why small eggs should be a good strategy from an evolutionary perspective, because being small also entails being exposed to a high mortality. Survival to adulthood was calculated earlier in eq. 8.2. For an adult size of 10 kg and egg size of 1 mg, the survival is $(10^4/10^{-3})^{0.58} \approx 0.0008$—the average offspring is a dead offspring. That low survival is of course offset by the higher number of small offspring that an adult can produce relative to large offspring. The adult reproductive output, measured in number over the adult lifetime, is the adult production $\varepsilon_{egg} A W_{\infty}^n / w_0$ multiplied by the expected adult lifetime, $1/\mu(W_{\infty})$. Here, ε_{egg} is the efficiency of egg production, which was found for teleosts to be 0.22 (fig. 3.5), and we saw in fig. 8.1c that it was not significantly different for elasmobranchs. Multiplying the survival and the adult reproductive output gives the lifetime reproductive output:[1]

$$R_0 = P_{w_0 \to W_{\infty}} \varepsilon_R \varepsilon_{egg} \frac{A W_{\infty}^n}{w_0} \frac{1}{a A W_{\infty}^{n-1}}, \tag{8.3}$$

$$= \frac{\varepsilon_R \varepsilon_{egg}}{a} \left(\frac{W_{\infty}}{w_0} \right)^{1-a}. \tag{8.4}$$

Note that I have further discounted reproductive success with the recruitment efficiency ε_R (see eq. 4.35).

Under reasonably general conditions, R_0 can be used as a fitness proxy (a population in equilibrium and density dependence happening as a multiplicative factor before maturation; Mylius and Diekmann, 1995). Under those conditions, the most succesful strategy is therefore one that maximizes R_0. In that context, the calculation of the lifetime reproductive output therefore has three implications, as follows:

1. Fitness R_0 scales with asymptotic size as $\propto W_{\infty}^{1-a}$. Fitness is therefore increasing with W_{∞} because $a < 1$—it is good to be large.

---

[1] A more elaborate calculation of R_0 can be obtained by combining the calculations in chapter 4: eq. 4.39 with eq. 4.35 and eq. 4.33. That formulation of R_0 will to some degree also represent that adults continue to grow (indeterminate growth). The result is the same up to a constant (Andersen et al., 2008).

2. Fitness scales with offspring size as $\propto w_0^{a-1}$. Fitness therefore increases if offspring size is decreased—it is good to make eggs that are as small as possible.

3. If eggs become too large relative to the asymptotic size (larger than $(a/(\varepsilon_R \varepsilon_{\text{egg}}))^{1/(a-1)} W_\infty \approx 0.0007 W_\infty)$, then $R_0 < 1$ and the population is no longer viable.

That fitness increases with asymptotic size was found earlier in eq. 4.39). However, the key result for understanding the success of the teleost offspring size strategy is the second one. Making eggs as small as possible, irrespective of adult size, is exactly what they do—1 mg appears to be the smallest possible size they can make.

That immediately raises another question: what limits the lower size of fish offspring—why don't they make even smaller eggs? After all, there is no general physiological limitation toward making smaller eggs. Copepods, for example, make eggs that can be several orders of magnitude smaller (Neuheimer et al., 2015). Two explanations can be brought forward. Fish larvae are visual predators. Organisms smaller than fish larvae, such as copepods, are blind and rely on tactile sensing to locate prey. This is because there is a lower limit to the size of a functional camera eye, at a diameter around 1 mm (Martens et al., 2015). The smallest visually foraging predator is therefore around 1 cm—exactly the size of a newly hatched fish larvae. Should fish make smaller offspring, they would have to devise another sensory organ than vision to aid their larvae feeding. Further, fish larvae feed using suction, but due to scaling of the hydromechanics at low Reynolds numbers, suction feeding becomes ineffective for fish larvae smaller than about 1 cm (China and Holzman, 2014). Taken together: were fish to make smaller eggs, the larvae would be blind, unable to feed efficiently, and consequently outcompeted by the tactile sensing copepods. The optimal offspring size strategy is therefore to make eggs or larvae that are around 1 cm, but not smaller.

8.4   WHY DO ELASMOBRANCHS MAKE LARGE OFFSPRING?

The advantage of making smaller offspring should in principle apply to elasmobranchs as well. Yet elasmobranchs do not make the smallest possible eggs, but make eggs proportional to their asymptotic size (fig. 8.2); elasmobranchs do not obey the prediction in eq. 8.3. In such a case, it is tempting to paraphrase the bumblebee explanation "scientists have shown that bumblebees cannot fly; they have just forgotten to tell them" as "elasmobranchs should make smaller offspring; evolution has just has been unable to figure out how." Evidently not a very satisfactory explanation. In the case of the bumblebees, science has since been convinced that bumblebees actually can fly,[2] and we should aim to find an explanation for the elasmobranch offspring size strategy as well. Let's start by taking a critical look at the theory developed earlier.

Simple optimization of R_0 has a fundamental flaw: it ignores the effects of density dependence and of competition between different offspring size strategies (also know as *frequency dependence*). It is a little like driving on the freeway and observing a traffic jam up ahead. We try to make the optimal decision of which lane moves fastest, and place ourselves there (the optimization). However, everybody else does the same, with the result that everybody ends up driving at roughly the same slow speed. In this way, the actions of everybody else shape the environment (the lane speed). Offspring size strategies are similar. Imagine a situation where a given growth and mortality schedule, such as the metabolic assumptions used earlier, results in an optimal small egg size. Everybody then follows that strategy, which leads to changes in growth and/or mortality due to density dependence. The changes in growth and mortality might alter the fitness such that the previously optimal strategy becomes suboptimal. Therefore, the environment, here growth and/or mortality, will change according to the strategy adopted by the majority of a population. To deal with this problem, we need to get density dependence explicitly into the theory.

For convenience, density dependence is incorporated into the equation of mortality (Charnov et al., 2013):

$$\mu = aAw^{n-1} \left(\frac{w}{W_\infty} \right)^{-d}, \tag{8.5}$$

leading to survival:

$$P_{w_0 \rightarrow W_\infty} = \exp \left[\frac{a}{d} \left(1 - \left(\frac{w_0}{W_\infty} \right)^{-d} \right) \right]. \tag{8.6}$$

The first part of this formulation of mortality is the same as the metabolic mortality used earlier (aAw^{n-1}). The second part imposes a steeper scaling of mortality toward maturation described by the parameter d: if $d > 0$, mortality decreases more with size than the metabolic predation. This increased mortality for smaller sizes can be interpreted as a result of density-dependent competition. Formulating the higher mortality as depending on the ratio w/W_∞ means that the effect of density

---

[2] The original calculation considered insect wings as airplane wings. The lift of a wing at different angles was well known from wind tunnel experiments, and finding the total lift was just a question of integrating over all the angles during the stroke of an insect wing. That calculation generates far too little lift to keep an insect hovering. The missing ingredient was the unsteady motion of the wings, which was elegantly taken into account by Jane Wang (2000). A beating wing creates a strong leading vortex that generates an order of magnitude higher lift, enough to make an insect fly.

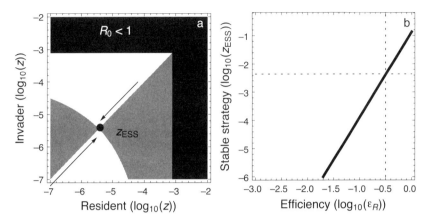

FIGURE 8.4. Offspring size strategies with explicit density dependence. (a) Pairwise invasibility plot showing the invasion fitness of an invading offspring size rate strategy $z = w_0 / W_\infty$ invading a resident offspring size strategy. Gray means positive invasion fitness, so the direction of evolution is towards the black point, which represents the evolutionary stable strategy, z_{ESS}. (b) The evolutionary stable strategy as a function of the reproductive efficiency ε_R. The horizontal dotted line is the observed offspring size strategy of elasmobranchs as determined in fig. 8.2, and the vertical dotted lines is then the equivalent reproductive efficiency, $\varepsilon_R \approx 0.3$.

dependence is strongest at small sizes, declining to no effect as juveniles approach adulthood.

The strength of density dependence is determined as the value of d leading to a population in equilibrium, $R_0 = 1$. As R_0 depends on the offspring size, so does d. With the strength of d defined, we have now defined a situation dominated by a "resident" offspring size strategy w_0, which determines the density-dependent mortality. We can now determine the evolutionary outcome of exposing this resident strategy to an invading strategy $\tilde{w}_0$ from the pair-wise invasibility plot fig. 8.4a. The plot shows that a small offspring size strategy will be replaced by a strategy with a larger offspring size, until the offspring size becomes about 1:100,000 of the adult size. That strategy is the *evolutionary stable strategy*, the ESS.

Note that the ESS depends only on the offspring:adult ratio, $z = w_0 / W_\infty$. This is because in the formula for survival and adult reproductive output w_0 and W_∞ always occur as a ratio. This insight is not only convenient, but it also shows that evolution will converge toward an offspring size strategy with offspring size being proportional to adult size—and this is just the strategy that elasmobranchs follow. The only problem now is that the offspring:adult ratio is far too small; 1:100,000 versus the observed ratio around 1:370 from fig. 8.2. Fig. 8.4b shows how increasing the recruitment efficiency ε_R produces larger offspring sizes. The recruitment efficiency that corresponds to the observed offspring size ratio is $\varepsilon_R \approx 0.3$. This is why I used $\varepsilon_R = 0.3$ for elasmobranch in the calculations of population growth rates in fig. 8.3.

By adding density dependence to our simple theory, we have shown how the proportional strategy is indeed favored by evolution. Further, we have used the theory to provide an indication of the value of the one life-history parameter that we were unable to compare between teleosts and elasmobranchs: the recruitment efficiency. For teleosts, $\varepsilon_R \approx 0.03$, while ε_R appears to be around a factor of 10 larger for elasmobranchs. Having the theory explain the elasmobranch offspring size strategy, however, invalidates our previous simple description for why teleosts make small offspring. We showed that the proportional strategy can occur if density dependence, which increases mortality or decreases growth, happens predominantly early in life. We believe that density dependence occurs predominantly early in teleosts (see section 4.3), so why do they not follow the proportional offspring size strategy? It turns out that if there is an initial larval period without density dependence, and if this period is sufficiently long, the small egg strategy will still prevail. I will demonstrate this result by combining the two theories for offspring size developed so far.

Combining the two theories developed earlier makes it possible to explain how two similar groups of organisms, differing only in their reproductive efficiencies, can have two distinctly different offspring size strategies (Olsson et al., 2016). The idea for the combined theory comes from a highly original paper about seed size strategies in plants by Falster et al. (2008). Falster and co-workers assumed that density dependence in plant growth occurs in three phases: (1) an initial phase occurs without competition between seedlings; (2) when the small seedlings reach a certain height, they begin to shade one another, and competition for light begins; and (3) adults do not grow any longer and density-dependent competition between equal-size adults ceases. The initial density-independent phase is crucial. Without that phase, the result of the preceding analysis where density dependence drove an ESS with large offspring will still hold. The theory considers a resident strategy for egg size w_0 that is being invaded by another strategy $\tilde{w}_0$. By the time the offspring following the resident strategy have reached the size where they begin competition, at size w_s, the invaders have reached another size $\tilde{w}_s$. The difference in size at the onset of competition may or may not give the invaders an advantage. I develop this idea in box 8.1 for fish along the same lines, just replacing the formulations of growth and mortality with the ones relevant for fish.

Fig. 8.5 shows pair-wise invasibility plots for the combined theory. The top row shows the results for a small reproductive efficiency, relevant for teleosts. Here, the small offspring size strategy is the only ESS, except at large asymptotic sizes where there are signs of a proportional strategy emerging. For higher reproductive

---

BOX 8.1

COMBINED OFFSPRING-SIZE THEORIES

Survival to maturity is divided into two phases: from offspring size to the transition size $w_s \approx 1$ g, and from w_s to adult size W_∞. The lifetime reproductive output is then

$$R_0 = P_{w_0 \to w_s} P_{w_s \to W_\infty} \varepsilon_R R_{\text{adult}}. \tag{8.7}$$

Survival in the first phase is density independent and given by eq. 8.4. Survival in the second phase is density dependent and given by eq. 8.6, just with w_0 replaced by w_s. As before, d is determined by the condition that $R_0 = 1$. We now introduce an invader with a different strategy $\tilde{w}_0$. The fitness of that strategy is

$$\tilde{R}_0 = P_{\tilde{w}_0 \to \tilde{w}_s} P_{\tilde{w}_s \to W_\infty} \varepsilon_R R_{\text{adult}}. \tag{8.8}$$

Note that the invaders' fitness depends on the invaders' weight, $\tilde{w}_s$, at the time when residents have reached the size where competition emerges. The time it takes residents to reach w_s is found by solving the growth equation for them, $dw/dt = Aw^n$ (eq. 3.24), to find

$$t_s = \frac{1}{A} \frac{w_0^{1-n} - w_s^{1-n}}{n-1}. \tag{8.9}$$

In that time period, the invaders reach a size

$$\tilde{w}_s = (w_s^{1-n} + \tilde{w}_0^{1-n} - w_0^{1-n})^{1/(1-n)}. \tag{8.10}$$

With that relation, we can determine the fitness of an invader from eq. 8.7, using $R_{\text{adult}} = \varepsilon_{\text{egg}} W_\infty / (w_0 a)$ from eq. 8.3.

---

efficiencies, relevant for elasmobranchs, the proportional strategy solidly emerges as the dominant ESS. The only exception is small asymptotic sizes; around $W_\infty = 100$ g the proportional strategy ESS is very weak, and it will disappear entirely for smaller asymptotic sizes (not shown). In this way, the combined theory explains both the teleost small offspring size strategy and the elasmobranch proportional strategy, as well as the absence of small elasmobranchs. The condition for the small offspring strategy to emerge is that the reproductive efficiency is low, and that there exists an early period in life for small offspring where survival is density independent.

This it not the most solid theory science has seen, but it is all we have for now. The theory is based on a fairly weak conjecture of a fundamental difference in

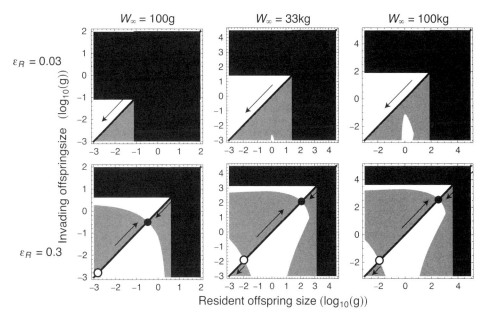

FIGURE 8.5. Pairwise invasibility plots for the theory with early-life density independence (box 8.1). The top row is with reproductive efficiency $\varepsilon_R = 0.03$ (teleosts) and the bottom row is with $\varepsilon_R = 0.3$ (elasmobranchs). The solid circles are convergence-stable ESS, the open circles are unstable ESSs, and the arrows show the direction of evolution. The plots show how evolution will progress toward either the smallest possible sizes (top row, low recruitment efficiency), or the coexistence of an ESS and a small offspring size strategy (bottom row).

the initial density-dependent control between teleosts and elasmobranchs. Until something more convincing comes up, the theory serves us well as a working hypothesis. At least, the theory explains all our observations: First, it accounts for the existence of a robust ESS at a size proportional to the asymptotic size. Second, it also predicts the existence of another ESS at egg size zero. Third, the theory predicts that the proportional strategy is more favored when the reproductive efficiency ε_R is high and vice versa. Further, the proportional strategy gets increasingly fragile as asymptotic size decreases and seems to disappear entirely around $W_\infty \approx 100g$, which corresponds to the asymptotic size of the smallest elasmobranch. Last, it is worthwhile to notice how the theory predicts a difference in the amount of density dependence needed to maintain each of the ESSs at population equilibrium: the teleost ESS needs much more density-dependent regulation than the proportional ESS. This difference is exactly what explains why teleosts are much more resilient to fishing than elasmobranchs.

8.5   SUMMARY

Despite being physiologically quite different, elasmobranchs and teleosts turned out to have similar values of most life-history parameters. The main differences relevant for demography are in the offspring size strategy and the associated difference in recruitment efficiency: teleost make many small offspring with a high initial loss, while elasmobranchs make offspring 1:100 smaller than their adult size with a low initial loss.

The key to the remarkable success of teleosts is the small offspring strategy. This strategy means that cohort biomass per recruit increases much more from offspring to adult size than it does for the elasmobranchs. A higher cohort biomass per recruit means that the population has a higher density-dependent regulation, which buffers against external perturbations. Therefore, teleosts are particularly resilient towards fishing.

The success of the small-offspring strategy makes one wonder why that strategy is not universally adopted by animals. Neuheimer et al. (2015) looked at offspring size strategies of all kinds of marine animals, and they found that the majority follow the proportional strategy with a roughly 1:100 offspring:adult size ratio: crustaceans, elasmobranchs, and marine mammals do it. What keeps these groups from developing even smaller offspring, now that it appears to be advantageous? The evolutionary theory provided two hints: for the small offspring size strategy to be evolutionary stable requires a period very early in life without density-dependent growth or mortality. Further, the small offspring strategy cannot be reached by incremental evolution because the proportional strategy is associated with a local ESS. Evolution therefore needs to make a large decrease in offspring size to jump from the proportional strategy ESS to the small offspring ESS. An alternative evolutionary explanation is that the small offspring ESS first emerges among species with small asymptotic size, where the proportional ESS does not exist. Once the small offspring strategy, with associated low ε_R, is established, it can radiate with incremental evolution to larger asymptotic sizes. Whether this is what really occurred may be visible from the phylogenetic record.

Do teleosts really have a density-independent phase early in life which is needed to make their small offspring size evolutionary stable? In section 4.2, I described how density dependence in fish is generally thought to occur early in life such that it can be described with a stock-recruitment relation. It is indeed correct that there is a strong density-dependent component early in life. It seems, however, as if density dependence does not happen right as the eggs hatch. Rather, density dependence appears to affect slightly older juveniles (Hixon and Jones, 2005), and be strongest post-settlement for demersal fish (Ford and Swearer, 2013). Later, I will show simulations of the entire fish community, in fig. 11.4. There, the abundance of newly hatched larvae is very small compared to the community spectrum, and only at a size around 1 g do they become the dominant group. This is the size where I expect density dependence to begin. That size is still early in life, but it also allows for a density-independent phase in the first month of the life of teleost fish. The various pieces of evidence, empirical and theoretical, therefore indicate that teleost fish indeed have a period of density-independent growth and mortality in their very early life, which is required for the evolutionary explanation of why teleost fish make small eggs.

CHAPTER NINE

Trait-Based Approach to Fish Ecology

Fish are one of the most diverse groups of vertebrates, with around 30,000 known species. They come in a myriad of sizes, shapes, colors, and life-history strategies: some are small, some are big; some are round, some are flat; some roam the productive pelagic zone, while others eke out a living in the deep sea on the detritus from the feast in the surface; some are active hunters, while others wait for the unaware prey to pass by. How can one obtain general insights about the demography and dynamics of such a rich and variable group of species? Throughout this book, I have developed the idea that focusing on a few well-chosen traits captures the most important variations among fish populations. I have highlighted the asymptotic size, W_∞, as the central trait, but mentioned other traits in passing—notably, the growth rate A (chapter 3) and the offspring size (chapter 8). However, I have yet to develop a definition of what constitutes a trait, explain how one selects the most important trait(s), and show how traits relate to classic life-history theory.

What is a *trait*? The ecological literature has swelled with position papers about various aspects of a trait-based ecology. The papers have developed concepts of *environmental filters*, listed all possible traits of groups of species, or tried to categorize traits into abstract groups such as *effect* and *response* traits. All that is probably very good, but here I will take a more pragmatic and concrete approach. A trait is any quantity that characterizes an individual organism. It should preferably be a directly measurable property, such as the shape of the tail or the size of the gonads, but it does not *have* to be directly measurable to be useful. The asymptotic size, for example, is not directly measurable, it appears as a parameter only when size-at-age is fitted to a growth equation. The definition of what constitutes a trait is therefore fairly loose. Note that traits characterize individuals; a population-level measure, such as the population growth rate, is not considered a trait. The most important quality about a trait is that it can be quantified, and that it carries as much information about the organism as possible.

Having defined a trait, the next question is which traits are most important. One can envision a high-dimensional trait-space spanned by axes representing all possible traits. Any species, any population, or any individual, is characterized by countless traits, each defining a single point in that trait-space. The crux of the trait-based approach is to project the high-dimensional trait-space onto a low-dimensional space. In other words, to select a few, or even just one, trait that characterizes most of the observed variation. Typically, many traits are correlated, and by factoring out these correlations the dimension of the trait-space is reduced. Such a reduction can be done formally with a principal component analysis. However, even when correlated traits are factored out, there are still many traits left, and the task of selecting the most useful traits, often referred to as *functional traits*, remains.

The most famous "master" functional trait is body size, be it length, diameter, or weight (Andersen et al., 2015). Body size correlates with almost all aspects of an individual's physique and its interactions with other organisms. If we pick a random living organism and are allowed to ask for just one parameter to describe it, we get the most information from its size. While body size describes an individual well, it is not a good description of a fish species or fish population, because body size varies so uncommonly much from the tiny 1 mg egg of a (teleost) fish to the asymptotic size that ranges from 1 gram to 1 ton. For a fish, body size is a state in life, rather than a trait that characterizes the species it belong to. Nevertheless, body size of individuals can still be very useful as a master trait for understanding community ecology. The celebrated *metabolic theory* of ecology (box 4.1) is essentially a trait-based theory based upon the premise that everything is determined by the metabolic capacity of individuals, and that metabolism is tightly linked to body size. Metabolic theory therefore works with body size as a master trait, but it uses body size to characterize only one aspect of physiology. The size spectrum theory in chapter 2 was also developed solely on the basis of body size; however, size spectrum theory went further than metabolic theory by also accounting for other aspects of an organisms physique, notably the clearance rate.

To go beyond body size, we can look for inspiration from the pioneers of the trait-based approach: the plant ecologists. Grime (1977) developed the idea of a plant's strategy as a point in a three-dimensional space spanned by the three axes of allocation to growth, reproduction, and maintenance. Since these allocations trade off against one another, the investment in one has to be at the expense of the two others. Therefore, the effective trait space is two-dimensional and a plant's strategy is a point inside a triangle. *Grime's triangle* stands central in the creation of trait-based ideas in plant ecology. He related the allocations to three central strategies: competitors, stress tolerators, and ruderals (first colonizers). Grime's triangle has had immense importance for organizing thinking about plants—the paper is cited more than 4,000 times. Nevertheless, the concept has been difficult to operationalize; Where exactly within the triangle should we place a given species? How do we quantify a plant's allocation between growth, reproduction, and maintenance? Doing that requires a direct association between the allocation and concrete measureable traits. Creating that link turned out to be harder than it seemed, and progress stalled.

In comes the concept of *functional traits*. Instead of repairing Grime's conceptual framework by tying the allocations up to measurable traits, plant ecologists started from scratch by defining the handful of traits that most characterize the variation in strategies among plants. Out of that exercise emerged a shortlist of plant traits (Westoby et al., 2002): (1) the ratio between the weight and the area of a leaf (the *leaf mass per area*), which quantifies the range between efficient and cheap leaves with a short life span (low leaf mass per area) to sturdy but less efficient leaves or needles with a long life span (high leaf mass per area); (2) seed size; and (3) height at maturation. These traits are easy to measure and collect for all species, and one can examine how they are expressed in different geographical locations or during a succession of vegetation. The definition of a set of functional traits stepped away from a description of specific species toward a simple understanding of global patterns of vegetation. By their example, plant ecologists have shown how functional traits can be wielded to operationalize an abstract description of life-history strategies. In this chapter, I will try to follow in the footsteps of plant ecologists by proposing a shortlist of fish "master" traits and connect these traits to classic life-history strategy thinking. First, I will take a step back to set the historical background for the current state-of-the-art thinking about fish life history strategies.

9.1   LIFE-HISTORY STRATEGIES

The first and most influential classification of life-history strategies is between r and K strategies (MacArthur and Levins, 1967; Pianka, 1970). These early works do not actually talk about life-history strategies, but rather about r- and K-selection with reference to the environment selecting for organisms with different strategies. The r-selection refers to environments that select for organisms with rapid population growth—a high value of the maximum population growth rate r_{max}. Species being r-selected are similar to Grime's ruderals; their strategy is to grow fast and reproduce fast but invest little in competitive ability and defence. These r-strategists will dominate early in a succession or in highly variable environments where their fast growth rates make them appear early. The contrast is the K-strategists who invest in competition and/or defense and make fewer but fitter offspring. This strategy enables them to dominate in saturated and stable environments with fierce competition for resources or a high predation pressure.

Pianka (1970) argued that large organisms are K-selected because of their low r_{max}, while smaller organisms are r-selected. Based on this argument, he concluded that fish are a special group because "Fish, in particular, span the range of the r-K continuum." In other words: small forage fish are extreme r-strategists, while large predatory fish are extreme K-strategists. This view is pervasive today; however, it is misleading. There are two mistakes in Piankas logic: First, the assumption that r_{max} declines with body size. We saw in chapter 7 that this was not the case; medium-size fish are expected to have the highest population growth rates (fig. 7.1). The other mistake is the implicit assumption that an organism competes with all other organisms—that is, that all organisms fill the same ecological niche. That is not the case: an elephant does not fill the same niche as a mouse. The comparison should rather be done between species within the same general niche, such as between plants, who all compete for the light, nutrients, and water regardless of their size.

Do all fish belong to the same niche? We can view the ecological niche of fish in two ways. They all rely on copepods, either during their entire life (forage fish), or at least during the larvae and early juvenile stages. By defining their niche from their reliance on copepods, it follows that fish do compete and that they fill the same niche. From this definition, the r-selected fish species are those with the higher population growth rate r_{max}. However, we could also focus on the adult stage of fish populations, as this stage has the largest biomass and therefore the biggest impact on the biotic environment (see fig. 4.2b). Adults of forage fish and large predatory fish do not fill the same niche because they feed on different sizes of food. They are not competitors; they are rather prey and predators. Using adults to define the niche, we should then look for r- and K-selected species only among species with similar asymptotic body size. If we do that, the growth coefficient, A, is a better a proxy for r- and K-selection: high growth rate means faster body growth, higher reproductive rate, and faster population growth rate, and vice versa. Whatever way you do the niche partitioning, it is clear that while the r/K dichotomy is a useful way to compare life-history strategies between fish, its apparent simplicity is deceptive and it is hard to tie it unambiguously down to the variation of a single trait.

As a way of breaking loose from the simple r/K thinking, Winemiller and Rose (1992) categorized teleost fish strategies by their reproductive traits. They identified two main trait axes: asymptotic length (correlates with length at maturation and large clutches) and egg size. On this basis, they identified three strategies: (1) The periodic strategy, with fish "which delay maturation in order to attain a size sufficient for production of a large clutch and adult survival during periods of suboptimal environmental conditions." Large capital breeding fish such as large gadoids and scombroids are typical followers of the periodic

strategy. (2) The opportunistic strategy is characterized by "early maturation, fre-
quent reproduction over an extended spawning season, rapid larval growth, and
rapid population turnover rates, all leading to a large intrinsic rate of population
increase." The opportunistic strategy is typical of small income-breeding forage
fish. (3) Last, they described the equilibrium strategy: "Large eggs and parental
care result in the production of relatively small clutches of larger or more advanced
juveniles at the onset of independent life." That strategy fits the elasmobranch
described in chapter 8. There is a strong similarity to the r/K selection here: the
opportunistic strategy is largely the r-strategy, while the periodic and the equilibrium strategies essentially divide the K-strategy in two: one for large and one for
small species.

9.2   TRAITS AND TRADE-OFFS

Winemiller and Rose's analysis of reproductive traits built a bridge between r/K
and traits. However, the analysis focused only on reproduction. A complete analysis of functional traits needs to consider that evolutionary success is not only about
reproduction; reproduction is preconditioned on growth and survival. I propose
that the main axes of variation between fish species can be captured by three traits:
the asymptotic size, W_∞; the growth rate coefficient, A; and the adult-offspring
mass ratio W_∞/w_0 strategy (fig. 9.1). Together, these three traits determine
the central demographic parameters: somatic growth rate, investment in reproduction, age at maturation, survival to maturation, mortality, and so on, and from
there follows population-level quantities like population growth rate, population
structure, fitness, and selection responses. The two first traits, asymptotic size and
growth rate, are continuous traits with a large variation: the asymptotic size varies
from around 1 g to 1 ton, while the growth rate varies more than a factor of 10. In
contrast, the adult-offspring mass ratio is a discrete trait that distinguishes between
the teleost and elasmobranch strategies: teleosts have a high adult-offspring mass
ratio that is independent of asymptotic size, while elasmobranchs make offspring
that are proportional to the adult mass (see fig. 8.2). This difference in offspring
size strategy is illustrated with round versus triangular symbols in fig. 9.1.

The most important population-level measure is the maximum population
growth rate, r_{max}. The maximum growth rate is indicated in fig. 9.1 by the size
of the symbols. Clearly, the faster the somatic growth rate A, the faster the population growth rate. However, in contrast to Pianka's take on the r/K theory, the
smallest fish do not have the fastest population growth rates. The species with
the fastest population growth rates in the data set is the mahi-mahi or dolphinfish
(*Coryphaena hippurus*). It is a large-bodied species living offshore in temperate,

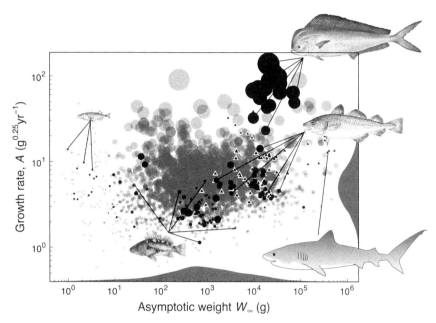

FIGURE 9.1. The trait-space of fish spanned by asymptotic size W_∞, growth rate coefficient A, and reproductive strategy: small eggs (teleosts; circles) and egg size proportional to adult size (elasmobranchs; triangles). The size of the symbols is scaled by the population growth rate as calculated from the approximation in eq. 7.5, with reproductive efficiency $\varepsilon_R = 0.03$ for teleosts and 0.3 for elasmobranchs (see section 8.1). I have singled out some species represented by several points (for example, Atlantic cod), and others that represent different species in the same family (dolphinfish, rockfish, and sticklebacks). Some stocks of small asymptotic size have negative growth rates, due to the inaccuracy of the analytic approximation of r. These stocks are represented by a small symbol. The two dark gray patches show the distributions of asymptotic sizes (on the x-axis) and growth coefficients (on the y-axis) for teleosts. Data from FishBase (Froese and Pauly, 2017).

tropical, and subtropical waters. Dolphinfish grow exceedingly fast, mature early in their first year, but do not live very long compared to other large-bodied species. It is easy to focus on the extreme cases, like the dolphinfish. However, their somatic growth rates are more than two standard deviations away from the mean growth rate, so they should really be regarded as an anomaly. Among the species with more moderate above-average individual and population growth rates, we find commercially important species such as Atlantic cod. Despite having a fairly large asymptotic size, cod also have fast somatic growth rates, and therefore fast population growth rates. The fast growth rate is what makes them such important and productive fish stocks. On the other end of the population growth rate spectrum are slow-growing teleosts, such as rockfish, with somatic growth rates almost a factor of 5 smaller than cod and more than a factor of 10 slower than dolphinfish. The slow population growth rate is what makes rockfish particularly sensitive to fishing. Among the smallest fish, I have singled out the tiny three-spined stickleback. Stickleback are common throughout the northern hemisphere, in fresh- as well as saltwater. Sticklebacks have fairly average somatic growth rates, but due to their small asymptotic size their population growth rates are smaller than larger species with similar somatic growth rates. Despite being ubiquitous in the northern hemisphere, sticklebacks are represented with only three data points—databases such as FishBase tend to be dominated by observations of species that have been singled out for studies due to exploitation (like cod), conservation (like rockfish), or because they are charismatic (like dolphinfish). The smallest population growth rates are found among the large elasmobranchs—I have singled out the tiger shark (*Galeocerdo cuvier*) as an example of a large shark that is found offshore in temperate and tropical seas worldwide. Clearly, the population growth rates of the elasmobranchs are much lower than the similar-bodied teleosts, which is why they are less common and more vulnerable to fishing than the teleosts.

The trait-based framework operationalizes the description of life-history strategies. In fig. 9.1, the r/K dichotomy appears as a pattern of r_{max} in trait-space: the r-selected species are those with high somatic growth rates and medium asymptotic body sizes. This insight adds more flavor to the r/K selection than just body size. The wide distribution of species in trait-space shows that one should be careful about grouping species into just a few categories—the trait-space of W_∞ and A is a continuum where species have carved out niches with any combination of trait values. If we should make one rough division between r and K strategists, it could be between the teleosts and the elasmobranchs. Winemiller and Rose did not cover elasmobranchs, but instead identified a third life-history strategy as the *equilibrium* strategy. The equilibrium strategy is adopted by elasmobranchs with a low fecundity. It is also adopted by small species that make relatively large offspring or have parental care. In trait-space, the smallest species are indeed somewhat special as they have relatively low population growth rates. It is among those small species we find the most diverse offspring strategies. While the smallest species do not appear to have systematically larger offspring than the large species (see fig. 8.2), they are more likely to make live offspring or engage in some form of parental care than the larger species. The male stickleback, for example, guards his eggs and takes care of newly hatched larvae.

The trait-based description of the diversity of species is convenient because it connects strategies with concrete measurable traits. In that aspect, it is not fundamentally different from the life-history descriptions. The life-history descriptions also developed continuous life-history spaces—for example, Winemiller and Rose's triangle of strategies—and they also developed a coarse link between traits and strategies. Where the trait-based description excels is in its ability to scale directly up to population- and community-level measures, such as the population growth rate and the Sheldon spectrum. That link is forged by developing the mechanistic basis of traits and trade-offs (Kiørboe et al., 2018).

The crux of the mechanistic trait-based framework is actually not the traits—it is really about the trade-offs. Trade-offs define the fences around the playground for evolution. Ideally, they are "hard" constraints rooted in conservation laws, physiology, and physics. A conservation trade-off could be allocation of energy between two processess, such as between growth or reproduction. Such allocation trade-offs were essential to Grime's thinking about plants that have to balance allocation between growth, reproduction, and maintenance. However, Grime's trade-offs were abstract and not easily connected to measurable traits. In fish, an allocation trade-off emerges between growth and reproduction (see section 3.2). A physiological limitation could be the efficiency and speed of a digestive system or of the gills of a given size. An example of a trade-off rooted in physics is the costs of movement due to fluid drag.

In the demographic model developed in chapters 3 and 4, the traits appear in several of the fundamental assumptions about fish physiology, as shown here:

$$
\begin{array}{lll}
\text{Available energy} & E_a = \overbrace{\varepsilon_a h(f_0 - f_c)}^{A} w^n & \text{(see eq. 3.30)} \\
\text{Mortality} & \mu = aAw^{n-1} & \text{(see eq. 4.7)} \\
\text{Size at maturation} & w_{mat} \propto W_\infty & \text{(see fig. 3.4)} \\
\text{Fecundity (no./time)} & \propto AW_\infty^n / w_0 & \text{(see eq. 3.19)}
\end{array}
$$

Generally speaking, trade-offs are related to the three central missions of life: to eat, to survive, and to reproduce. The trade-offs quantify costs and benefits. The benefits of a large asymptotic size is lower adult mortality and a larger fecundity. These obvious benefits come at the cost of later maturation. A high growth-rate coefficient A has the immediate benefits of faster growth and higher fecundity. However, fast growth also comes at the costs of a higher demand for food and a higher standard metabolism $hf_c w^n$. Further, the increased activity associated with higher consumption leads to an elevated mortality—fast living comes at the risk of dying young. A smaller offspring size leads to more offspring, but smaller offspring also means that it takes longer to mature and exposes the offspring to a high mortality. More insight into the benefits and costs comes when the fundamental assumptions about trade-off are scaled up to population-level measures, as shown here:

$$
\begin{array}{lll}
\text{Survival} & P_{w_0 \to W_\infty} = \left(\dfrac{W_\infty}{w_0}\right)^{a-1} & \text{(see eq. 4.31)} \\
\text{Adult life-time fecundity} & = \dfrac{1}{a}\dfrac{W_\infty}{w_0} & \text{(see eq. 8.3)} \\
\text{Population growth rate} & r_{max} = AW_\infty^{n-1} f(W_\infty/w_0) & \text{(see eq. 7.13)}
\end{array}
$$

Survival is a declining function of W_∞/w_0 because $a < 1$. Therefore, increasing asymptotic size leads to lower survival to adulthood. Yet, higher W_∞ also leads to increased adult lifetime fecundity and faster population growth rates. The growth rate coefficient A is revealed to be almost neutral—faster growth does not translate into higher survival or higher adult fecundity. This is because of the trade-off with mortality risk that exactly balances the gains of faster growth with corresponding increased mortality. The only place where the growth rate coefficient appears is in the population growth rate. Survival and adult fecundity show the importance of the adult-offspring size ratio W_∞/w_0. In practice, this difference is played out between two strategies: elasmobranchs follow the constant adult-offspring size ratio, while teleosts minimize w_0 (see chapter 8).

The choice of the fundamental traits axes that span trait-space involves some degree of arbitrariness. I could, for example, have chosen to select the size at maturation as the fundamental trait instead of the asymptotic size, or the coefficient of the maximum consumption rate h instead of the growth rate coefficient. I chose W_∞ and A simply by convenience: these two parameters follow more or less directly from the typically measured von Bertalanffy growth coefficients (with the procedure described in box 3.2), which are readily available for many fish species.

9.3   THE SWEET SPOT OF COMPLEXITY

At the core of any theoretical description of nature, be it the dead or the living, lies a reduction of information. The art of scientific theory is to throw away the less important stuff and retain only the really crucial information. Occam's razor describes the reduction thus: among competing hypotheses, the one with the simplest set of assumptions—the largest reduction of information—is superior. A modern version is the classic Einstein quote, "Everything should be made as simple as possible, but not simpler." The mechanistic trait-based approach is a concrete recipe for reduction of information, by projection onto a low dimensional trait-space.

Besides being an aesthetically appealing reduction of complexity, the trait-based approach lends itself directly to applications in three ways: description of stock dynamics (in chapter 5), data-poor stock assessment (Kokkalis et al., 2017), and community-level impacts of fishing (to be developed in chapter 12). However, there is more to fisheries advice than a handful of traits, and the trait-based approach will not supersede other approaches. We will still have single-species-based impact assessments and stock assessment that incorporate much more information than will be included in the pure trait-based descriptions. We want to know what happens exactly to, for example, the Coho salmon from the Columbia River, and getting that knowledge requires complex models of early life, the migration to sea and adult life, and the return to the native river, and so on, and the same for other stocks. We will also continue developing food-web models that resolve the specific species. We need such models to develop assessments of the impact of fishing that go beyond the general patterns revealed by trait-based models, such as knowing exactly how, for example, the sand eel fishery affects the cod fishery, or how the fishery affects the birds and mammals that feed on sand eel. These are hard questions that require a lot of effort, and we will continue to struggle for even more refined answers. In the meantime, the trait-based models are appealing compromises that provide easy quantitative predictions with a minimum of data and effort.

In this part, I have largely left fisheries applications behind to apply the theory to questions in evolutionary ecology of fish. The applications developed here came about naturally because of the close kinship between trait-based theory and evolutionary ecological theory. Evolutionary theory builds upon traits and trade-offs. As traits and trade-offs are also the core of the trait-based approach, it is clear that the mechanistic trait-based approach shares a conceptual basis to evolutionary ecological theory. However, the mechanistic trait-based approach goes further than evolutionary ecology: it can be used to calculate population structure (part II), and in part IV, I will develop the trait-based approach further and use it to describe the trait structure of fishing communities (chapter 11) and community-level responses to fishing (chapter 12). By being based on trade-offs, the mechanistic trait-based approach unites evolutionary ecology and population and community ecology. The conceptual leap that makes it possible to extend evolutionary ecology is the courage to forget about specific species and species-based food webs and instead focus on species-transcending traits and trait distributions. With trait distributions, trait-based theory manages to balance on Occam's razor, in the sweet spot between triviality and complexity.

Part IV
COMMUNITIES

CHAPTER TEN

Consumer-Resource Dynamics and Emergent Density Dependence

So far, the population modeling has followed the standard type of fish model used in fisheries science and management. The model described a population that lives in a constant environment with plenty of food and a mortality that is independent of its own abundance—that is, growth and mortality were fixed. Density dependence was imposed as a stock-recruitment relationship. The model is the antithesis of most population ecology, which is about resolving the interactions between a population and its environment—in particular, the interplay between the population and its resource and its predators. In this chapter, I extend the previous static model to become a fully dynamic consumer-resource model of the type we know and love from classic population ecology.

Classic theoretical population ecology describes the interactions between a consumer and a resource, or between prey and predators, or both. Essentially a consumer-resource model is the same as a predator-prey model, with the resource playing the role of prey and the consumer being the predator. A classic model is the Rosenzweig-MacArthur model (Rosenzweig and MacArthur, 1963), which is a Lotka-Volterra model with a carrying capacity on the prey and a type II functional response for the predators. To describe the interaction of organisms with different sizes, Yodzis and Innes (1992) introduced a scaling of the governing parameters with body size. Such unstructured models do a poor job of describing fish populations because they do not represent the extended size structure of fish, from the offspring around 1 mg to the asymptotic size. During their life from larvae to adults, fish feed on different prey, from large unicellular plankton and copepod nauplii, over adult copepods, and in many cases fish. We refer to such changes in food as *ontogenetic trophic niche shifts* (Werner and Gilliam, 1984). Most fish species have one or more ontogenetic trophic niche shifts, but the unstructured consumer-resource models describe predators with only one trophic niche. An adequate model needs to resolve the trophic niche shifts of fish.

Following the cue from Werner and Gilliam (1984), André de Roos and Lennart Persson developed *physiologically structured models* as a means to deal with the ontogenetic trophic niche shifts of predators. Those models scale from individual level processes up to describe the dynamics and demography of a population that feeds on one or more resources. Growth and reproduction is described by an energy budget similar to the one developed in chapter 3. Density dependence is not prescribed by a stock-recruitment relation but emerges from competition for resources, control by predators, and/or cannibalism. Physiologically structured models show a rich type of dynamics, with stunted growth, cohort cycle dynamics, and multiple stable states, all of which are comprehensively documented by de Roos and Persson (2013). Many of those phenomena are governed by the competition for a single resource—for example, invertebrate prey—between differently sized individuals, and the outcome is determined by whether adults or juveniles are most competitive for the prey. Such situations where a single resource is key for the dynamics is typical of small lake systems with just one or two fish species. For these cases, physiologically structured models have been demonstrated to describe the observed complex dynamics well (Persson et al., 2007). Here, I consider an extended prey resource that covers all sizes of plankton and fish in the ocean. In this way, individuals can smoothly switch between differently sized resource items as they grow, and the model dynamics becomes less complex.

Why have fisheries science and management ignored consumer-resource dynamics in their models? The answer lies in a belief that density dependence in fish populations occurs only early in life combined with the necessity of a pragmatic approach to develop operational models. If density dependence happens predominantly early in life, then it is not really necessary to account for changes later in life due to competition for a resource or cannibalism. It is therefore safe to ignore consumer-resource dynamics and use the model with fixed growth and mortality that I developed in chapter 4. Since the work on the critical period early in life by Hjort 100 years ago (Hjort, 1914), fisheries science has operated with the idea that the strength of a cohort is determined by the survival in early life. This assumption was later embodied in the stock-recruitment relationship (see section 4.3). Despite the central role of early density dependence in fisheries models relatively little effort has been invested in understanding why density dependence should occur only early in life, and why not late in life—for example, through changes in adult growth.

Even though at least some density dependence most likely occurs early in life, we actually do observe density dependent changes in growth. A good example was offered by the pause in fishing in the North Sea during World War II (Graham, 1948), or the recent development in North Sea plaice (fig. 10.1). After the war, many stocks had recovered, and several stocks saw increased abundances accompanied by reductions in growth. Despite the evidence that density-dependent reduction in growth occurs, it is incorporated into fisheries advice

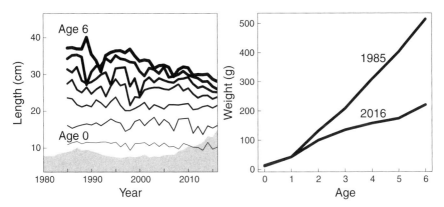

FIGURE 10.1. Size-at-age of North Sea plaice. The gray region shows the stock biomass. As the stock recovered in the recent decade, growth has declined noticeably. Data kindly provided by Tobias van Kooten (ICES, 2018).

only through observed weight-at-age curves that are used to calculate the spawning stock biomass from the age distribution. The reason for not incorporating density-dependent growth is pragmatic. First, many stocks have historically been overfished, and therefore density-dependent changes in growth were not occurring. Second, the information needed to confidently parameterize the effects of density-dependent changes in growth is just not operationally available. And even if it was available, the added benefit would in many cases be small compared to the huge uncertainty in other processes. Adding density-dependent changes in growth is like polishing the chrome on a racing bike—it makes the bike shiny, but not faster. This perception may change now that many fish stock are recovering and the attention of fisheries management shifts from reining in overfishing to managing sustainably fished stocks and ecosystems (van Gemert and Andersen, 2018a).

Both interpretations (early life density dependence and late density dependence emerging from competition) have problems. Fisheries management is not quite right in assuming that density dependence happens only early in life. On the other hand, assuming that density dependence acts only through processes late in life may lead to very strong responses, such as stunted growth and cohort cycles, which are rarely observed in marine systems (exceptions are the Baltic cod, which currently seems to show stunted growth [Eero et al., 2012] and the cohort cycles commonly observed in haddock stocks). In practice, all three effects of density dependence—early density dependence through limitations in the juvenile habitat, competition for resources, and cannibalism—act together to various degrees. This realization leads directly to the two questions that are the heart of the matter in this chapter: What is the balance between the three types of density dependence (early life, competition, and cannibalism) in fish populations in situ? And does it matter which type of density dependence is represented in a model of a fish population? We can extend the model from chapter 4 to address the second question, but the first question is harder.

In this chapter, I develop a physiologically structured consumer-resource model by extending the static model in chapter 4. I then study how density dependence emerges in the model, and how it changes the population size spectrum. Last, I explore how some of the standard fisheries impact assessments from chapter 5 are changed when density dependence is in the form of competition or cannibalism. Specifically, I show how the appearance of late-life density dependence rocks one of the cornerstones of contemporary fisheries management: that we should fish only the largest fish. In some cases, it turns out that yield is instead maximized by fishing juveniles.

10.1   A CONSUMER-RESOURCE MODEL

I base this model on the population model in chapter 4 and add the bits and pieces needed to make the model dynamic and describe the interaction with the resource. This entails combining the basic theory from earlier chapters: modeling of growth and reproduction from chapter 3, selection of food according to size from section 2.3, and mortality from section 2.5.

As in chapter 4, the model describes a single population. Before, the growth and mortality rates of the population were fixed. Now, growth and mortality are determined by the community spectrum (chapter 2), which acts as a resource and a source of predation mortality. The community spectrum diminishes as it is being eaten. Further mortality originates from cannibalism. In this way, the central processes, availability of food and mortality, are determined by the abundance—that is, they become density dependent.

Consumption

The available food is represented by the resource spectrum, N_{res}, and by smaller individuals of the species N (units of numbers per mass per volume). Food is selected by the size-based preference $\phi(w_p/w)$ as described in eq. 2.8 and fig. 2.7: a predator prefers prey of a size w_p that is a factor of β smaller than its own size. The available food E (mass per volume) is the integral over the resource and population spectra weighted by the preference for size

$$E(w) = \int_0^\infty (N_{res}(w_p) + N(w_p))w_p\phi(w_p/w)\,dw_p. \tag{10.1}$$

The actual encountered food, $E_e(w)$ (mass per time), is the available food multiplied by the clearance rate γw^q (volume per time; eq. 2.6)

$$E_e(w) = \gamma w^q E(w). \qquad (10.2)$$

Not all encountered food may be consumed. If the individual encounters more than it can maximally consume—that is, if $E_e(w) > hw^n$—it does not consume all that it encouters. The reduction in feeding level is described by a functional response type II

$$f(w) = \frac{E_e(w)}{E_e(w) + hw^n}, \qquad (10.3)$$

which varies between 0 and 1. The feeding level is the consumption relation to the maximum consumption rate, and the consumption is then $C(w)f(w)$. When little food is encountered ($E_e \ll hw^n$), then all encountered food is consumed and the feeding level $\approx E_e/(hw^n)$—that is, proportional to encountered food. When food is abundant ($E_e \gg hw^n$), the feeding level approaches 1, and the consumption approaches the maximum consumption rate.

Dynamic Energy Allocation

Chapter 3 developed a budget for how an individual fish uses the energy it acquires from consumption (section 3.3) for growth and reproduction

$$g(w) = E_a(w) \left[1 - \psi_m(w/w_m) \left(\frac{w}{W_\infty} \right)^{1-n} \right], \qquad (10.4)$$

$$R_{\text{egg}}(w) = \varepsilon_{\text{egg}} E_a(w) \psi_m(w/w_m) \left(\frac{w}{W_\infty} \right)^{1-n}, \qquad (10.5)$$

where ψ_m is the function that describes maturation around size $w_m = \eta_m W_\infty$, and E_a is the available energy

$$E_a(w) = \varepsilon_a(f(w) - f_c)hw^n. \qquad (10.6)$$

The function $f(w)$ is the feeding level developed earlier, which is the consumption relative to maximum consumption hw^n. Similarly, the critical feeding level f_c is the feeding needed to sustain standard metabolism and activity, again relative to maximum consumption. The remaining energy is assimilated with efficiency ε_a. Last, ε_{egg} is the conversion efficiency of food to eggs (see eq. 3.20).

The description in section 3.3 was a static energy budget where the feeding level was set to a constant, $f(w) = f_0$. Now the feeding level $f(w)$ is allowed to

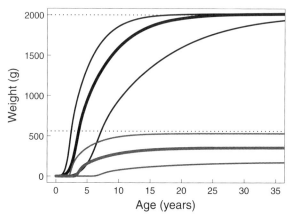

FIGURE 10.2. Weight-at-age (black lines) and egg production (gray lines) under different levels of food: $f = 0.4$ (lower lines), $f = f_0 = 0.6$ (thick lines), and $f = 0.8$ (upper lines). Asymptotic size $W_\infty = 2$ kg. Horizontal dotted lines are the asymptotic size and size at maturation.

vary with size and over time. Strictly speaking, I should write $f(w, t)$, and similarly elsewhere, to signal the time dependence, but I will generally omit t for brevity. The dynamic growth model eq. 10.4 has three characteristics: (1) It prioritizes evenly between growth and reproduction. If the feeding level declines, the individual grows slower and reproduces less. (2) Nevertheless, the individual will eventually reach the asymptotic size. (3) The only exception is cases of starvation where $f(w) \leq f_c$, where both growth and reproduction ceases. Fig. 10.2 shows how variation in feeding level from the expected level of $f = f_0$ gives rise to growth curves and reproduction that deviate from the static von Bertalanffy–like growth curves.

Reproduction and Recruitment

The production of new offspring (numbers per time) is determined by the total egg production, which is the integral over all individuals in the population

$$R_p = \varepsilon_R \int_{w_0}^{W_\infty} R_{\text{egg}}(w) N(w) \, dw, \tag{10.7}$$

where ε_R is the recruitment efficiency (section 4.2).

In principle, that is all we need to describe recruitment of new individuals in the population. However, I will also include a stock-recruitment relation that describes how early-life density dependence reduces the recruitment from the total egg production. I use the same *Beverton-Holt function* as earlier (see fig. 4.3 and eq. 4.36)

$$R = R_{max} \frac{R_p}{R_p + R_{max}}. \tag{10.8}$$

The stock-recruitment relation specifies the flux of recruits, R, at the size w_R. It is a saturating function of the egg production R_p limited by the maximum recruitment R_{max}. In practice, I will use the parameter R_{max} to adjust the amount of early-life density dependence that is imposed. In this way, I can switch smoothly from a recruitment that is purely determined by the egg production (large values of R_{max}) to purely determined by the stock-recruitment relation for small values of R_{max}.

Mortality

Mortality on the population arises from three sources: predation, cannibalism, and starvation.

Predation by other species arises from larger individuals in the resource spectrum. For simplicity, I here assume that the resource spectrum is always at its carrying capacity. That assumption has the advantage that we can use the solution to the predation derived earlier in eq. 2.15 to give

$$\mu_b(w) = \mu_0 w^{n-1}. \tag{10.9}$$

This equation essentially states that predation mortality declines with body size. The coefficient was derived earlier and related to the physiological mortality a as $\mu_0 = a\varepsilon_a(f_0 - f_c)h$ (eq. 4.41).

Cannibalism arises when larger individuals in a population eat smaller ones. The derivation involves an integral over all predators in the population multiplied by their prey size preference. The derivation is a little complex and is given in box 10.1.

Last, starvation occurs when the feeding level falls below the critical feeding level, when $f(w) < f_c$. Under those conditions, the individual has insufficient food to cover basal metabolic needs, and consequently it has no energy available to allocate to growth and reproduction. In the model, starvation results in an elevated mortality risk. I assume that starvation mortality is proportional to the energy deficiency $-E_a$ and inversely proportional to lipid reserves, which are assumed proportional to body mass

$$\mu_s(w) = \begin{cases} 0 & E_a(w) \geq 0 \\ -\frac{E_a(w)}{\xi w} & E_a(w) < 0 \end{cases}, \tag{10.10}$$

where ξ is a constant that sets the magnitude of the starvation mortality. In the simulations presented here, starvation does not play a big role; individuals may have varying feeding levels, but they rarely starve. In nature, starvation is an important process associated with *stunted growth*. Stunted growth happens when there is insufficient prey of a suitable size for adults in a population. In that situation, adults and juveniles compete for the same food resource, and depending on the relative competitive ability of juveniles and adults, individuals may just barely be able to grow to maturation before starvation occurs. Stunted growth is frequently observed in small lakes with just a single population of fish, such as roach, perch, or trout (Burrough and Kennedy, 1979; Ylikarjula et al., 1999) and is rarely observed in marine environments, probably because it usually disappears if a larger predator is present (Rask, 1983; Persson et al., 2007). Properly modeling stunted growth requires more attention to the process of starvation than I do here, which can be achieved by introduction of an extra state variable to account for reserves. A comprehensive modeling of stunted growth and the associated effects can be found in the book on physiologically structured population models by De Roos and Persson (2013). For the purposes of describing population dynamics of most marine populations, the simple description of starvation in eq. 10.10 suffices.

Resource Dynamics

The resource represents other individuals in the entire ecosystem of all sizes. I model the resource with chemostat dynamics

$$\frac{dN_{res}(w)}{dt} = r_0 w^{n-1} \left(\kappa_{res}(w) - N_{res}(w)\right) - \mu_p(w)N_{res}(w). \tag{10.11}$$

The chemostat dynamics is inspired by classic models of the upper water column, which have a similar dynamics (Evans and Parslow, 1985). The carrying capacity is described by the theoretical solution of the community size spectrum (eq. 2.20)

$$\kappa_{res}(w) = \kappa_{res0}w^{-2-q+n}, \tag{10.12}$$

and the population growth rate $r_0 w^{n-1}$ scales metabolically with size.

Last, the system is adjusted such that when the resource is at its carrying capacity, then the feeding level is at an reasonable level, $f(w) = f_0 \approx 0.6$. I do that by adjusting the coefficient for the clearance rate γ accordingly. Inserting $f = f_0$, $N_{res} = \kappa_{res0}w^{-2-q+n}$, and $N(w) = 0$ in the equation for the feeding level eq. 10.3, and isolating γ gives

$$\gamma = \frac{f_0 h \beta^{n-q}}{(1 - f_0)\sqrt{2\pi}\kappa_{res0}\sigma}. \tag{10.13}$$

We now have all the ingredients for the dynamic consumer-resource model. The full set of equations and parameters is given in appendix B. Parameter values are the same as for the static single-stock model in chapter 4 augmented with parameters that describe the dynamic allocation of energy (assimilation efficiency ε_a and critical feeding level f_c), and the resource dynamics (r_0 and κ_{res0}). The numerical solution procedure is the same as in chapter 7, described in boxes 7.2 and 7.3.

Is the Size-Based Consumer-Resource Model a Physiologically Structured Population Model?

A physiologically structured model describes the demography of a population structured and derived only from the state of individuals (Metz and Diekmann, 1986). The individual state is a vector that describes pertinent characteristics of individuals, such as age, size, condition, and so on. It should be possible to derive the demography from only the knowledge of the state of the individual and the environment. For the model derived in this section, the individual state is just the body size, and the environment is the resource. So far so good; however, according to the definition of a physiologically structured population model, dogmatically, the model is not a physiologically structured population model. The problem is the addition of the stock-recruitment relationship, which is a population-level process and not an individual-level process. However, if we disregard the stock-recruitment relationship, then the model is indeed a physiologically structured model. I use the stock-recruitment relation as a means of parameterizing those density-dependent effects early in life, which are not represented explicitly by the model. I will, however, also analyze the model without the stock-recruitment relation at all, and in those cases the model is indeed a bona fide physiologically structured population model.

The most important applications of physiologically structured population models have been through the work of André de Roos and Lennart Persson and co-workers in a long series of publications from the early 2000s and onward, where they systematically explored various resource-consumer-predator motifs. In the process, they uncovered several important processes: the relation between predator-prey and cohort cycles (de Roos and Persson, 2003), the emergent Allee effect (De Roos and Persson, 2002), the appearance of multiple stable states (Persson et al., 2007), and the process of biomass overcompensation whereby imposing a mortality may actually increase the biomass of a population (De Roos et al., 2007). Many of these effects are shaped by how differently sized individuals feed on the resource(s). One difference between those models and the one presented here is in the representation of the resource. Most often the resource is represented as one or two unstructured state variables characterized by one body size and consequently one trophic level. In the present model, the resource represents all body sizes of organisms in the sea and therefore represents many trophic levels. Working with this extended resource means that the shifts in feeding from one resource to another, or to piscivory, are less abrupt. Therefore, I have also not discovered any of the population dynamical effects mentioned above in this model (but see Hartvig and Andersen, 2013, for examples of emergent Allee effects and multiple stable states). That does not mean that those effects are wrong or that they do not appear in natural ecological systems. Those effects are certainly present in the cases where those models are the right representation of the system. However, it does indicate that some of the effects disappear or become less pronounced when the consumers have the option of feeding on numerous resources of different sizes.

10.2   EMERGENT DENSITY DEPENDENCE

The consumer-resource model expresses all three types of density dependence: early (through the stock-recruitment relation), competition (reduced growth), and cannibalism. The density dependence from the stock-recruitment relation is prescribed to act early in life, while the two other types of density dependence are not prescribed, but their effects emerge from the dynamics of the model.

Fig. 10.3 compares the output of the consumer-resource model (solid lines) with the demographic model from chapter 4 (dashed lines). The differences between the models, highlighted with shading, are the results of emergent density-dependence processes (competition and cannibalism). How and at which size density dependence operates is best seen in panel b, which shows the feeding level and the losses to mortality. The feeding level (the upper lines) decline for the largest individuals, which leads to reduced weight-at-age (panel d). The reduction in feeding occurs because the resource is depressed by the predation pressure imposed by the mature individuals with a size around 600 g. This reduction of the resource leads to density-dependent growth.

The peak in adult biomass also leads to increased cannibalistic mortality (panel c). The effect of cannibalism is manifested as a reduction in the size spectrum around a size of 1 g. The mortality losses in the population can be described the ratio between mortality and weight-specific maximum consumption

$$l(w) = \frac{\mu_b(w) + \mu_p(w)}{hw^{n-1}}. \tag{10.14}$$

The loss $l(w)$ indicates how much of the ingested food in a population is lost from the population through mortality. The loss is on average less than the feeding level. In fig. 10.3b, we see how the loss increases due to cannibalism. Comparing the gray areas of density-dependent feeding level and losses, we see that cannibalism imposes a bit more density dependence than competition in this example.

We can mix early density dependence with the emergent late density dependence by changing the value of the maximum recruitment R_{max}. If R_{max} is very

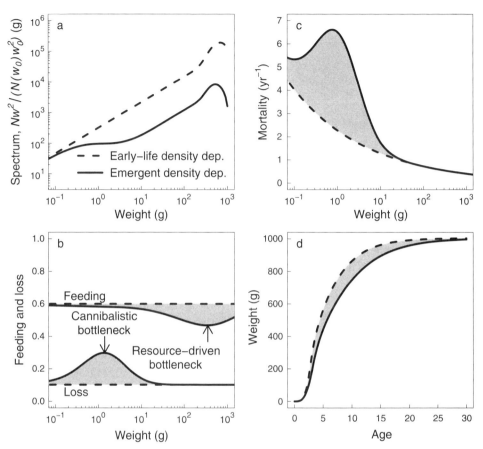

FIGURE 10.3. Comparison of results between the consumer-resource model (solid lines) and the model with fixed growth (dashed lines). (a) Biomass spectra; (b) mortality, with the gray area indicating cannibalism; (c) feeding level and loss, with the gray areas indicating lower feeding due to competition and increased loss due to cannibalism; and (d) weight at age.

high ($R_p \ll R_{\max}$), the recruitment is given by the egg production $R \approx R_p$ and the population will be entirely regulated by the emergent density dependence; if $R_{\max}$ is very low ($R_p \gg R_{\max}$) the recruitment is limited by the stock-recruitment relation, $R_p \approx R_{\max}$, and the population will be regulated entirely by the stock-recruitment relation. Now, what are "high" and "low" values of $R_{\max}$? In fig. 10.4, I vary the value of $R_{\max}$. As $R_{\max}$ is increased, the overall level of the size spectrum increases, until the point where the adults (the largest sizes) hit the resource spectrum. At that point, we begin to see reductions in growth and an increase of cannibalism. We can estimate the value of $R_{\max}$ where the spectrum hits the resource spectrum, as the value of $R_{\max}$ where $N(W_\infty) = N_{\mathrm{res}}(W_\infty)$, where the left-hand

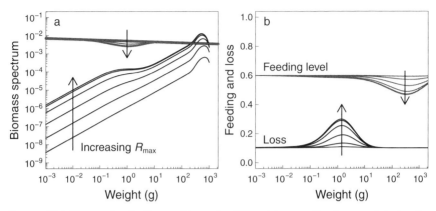

FIGURE 10.4. Size spectra, feeding, and losses for different degrees of early versus late density dependence. The arrows indicate increasing values of R_{max}, corresponding to increasing importance of emergent density dependence relative to imposed early density dependence. R_{max} ranges from $0.1 \ldots 1000 \times \tilde{R}_{max}$.

side is the population spectrum and the right-hand side is the resource spectrum. Approximating the species spectrum as $N(w) \approx R_{max}w^{-n-a}w_0^a/A$ (eq. 4.11) and assuming that the resource is at its carrying capacity (eq. 10.12), we can find the crossover R_{max} as

$$\tilde{R}_{max} = A\kappa_{res0}w_0^a W_\infty^{a+2n-q-2}. \tag{10.15}$$

Values of $R_{max} \gg \tilde{R}_{max}$ lead to dominance of emergent late density dependence and $R_{max} \ll \tilde{R}_{max}$ to dominance of early density dependence.

We now have an advanced consumer-resource model of the demography and dynamics of a fish population where we can manipulate the type of density dependence that rules the population dynamics by changing R_{max}. Essentially, if R_{max} is high (relative to $\tilde{R}_{max}$), then density dependence is ruled by the resource and cannibalism, and it emerges late in life. If R_{max} is small, then density dependence is ruled by the stock-recruitment relation imposed early in life. A more complete analysis of the model will show that we can also tune the density dependence between competition and cannibalism by changing the resource productivity r_0. That analysis is found in Andersen et al. (2017), and I will not go through it here.

The model can in principle be set up to simulate any given stock provided that we know the kind of density dependence that operates. For example, for the North Sea plaice stock in fig. 10.1, we know that there is indeed some early life density dependence due the limited size of the nursing habitat (Rijnsdorp and Leeuwen, 1996), but we can see that there is also late density dependence manifested as competition. We can then adjust the value of R_{max} until we reproduce the observed dynamics (van Gemert and Andersen, 2018b). However, closer inspection of

BOX 10.1

PREDATION MORTALITY

Predation mortality is calculated such that it exactly mirrors the amount of biomass consumed. In this way, the model balances the mass used for growth and reproduction of predators with a corresponding death of prey. Predators in the size range $[w : w + dw]$ consume $\phi(w_p, w)f(w)hw^n N(w)dw$ of w_p-sized prey. The total amount of food available from all prey to the predators in $[w : w + dw]$ is $E(w)$ (eq. 10.1). The mortality on prey of size w_p is the ratio between the actual consumption and the available food, integrated over all predators

$$\mu_p(w_p) = \int_0^\infty \frac{\phi(w_p/w)f(w)hw^n N(w)}{E(w)} dw. \qquad (10.16)$$

Using the formulation of the feeding level eq. 10.3, the mortality can be rewritten as

$$\mu_p(w_p) = \int_0^\infty \phi(w_p/w)(1 - f(w))\gamma w^q N(w) \, dw. \qquad (10.17)$$

The appearance of the term $1 - f(w)$ is not immediately obvious. It reflects the effect that a more satiated a predator (higher feeding level) eats a smaller fraction of the food it encounters.

fig. 10.1 shows that the pattern must be more complex. The recovery of the stock began in earnest after 2000; however, the decline of growth started before that. Therefore, reasons other than the increasing stock size for the decline in growth must be responsible for the declining growth—for example, a lower productivity of the resource, or competition with other species. Despite the difficulties of accurately understanding and mimicking the type of density dependence in nature, we can still use the model to study how different types of density dependence affect a stock's structure, dynamics, and response to fishing.

10.3   WHEN IN LIFE DOES DENSITY DEPENDENCE OCCUR?

Before approaching the question of when in life density dependence occurs, let us recap what is meant by *density dependence*. Density-dependent processes are those where the values of vital rates, such as growth, mortality, or reproduction, vary according to the density (abundance per area or volume) of the population. A more intuitive description could be "crowding effects." Examples of density dependent effects are lowered growth among adults (Lorenzen and Enberg, 2002) due to competition for food; stunted growth and consequential lowered reproduction in small lakes, also due to food competition (Burrough and Kennedy, 1979; Ylikarjula et al., 1999); cannibalism (Smith and Reay, 1991); or prey switching by predators (Myers and Cadigan, 1993). Of course, these density-dependent effects are not mutually exclusive but may occur in concert (Myers and Cadigan, 1993): there could be a density-dependent bottleneck among juveniles competing for a juvenile habitat in combination with competition for food among adults, as observed for plaice in the North Sea (van der Veer, 1986; Rijnsdorp and Leeuwen, 1992).

The consumer-resource model showed that density-dependent processes of competition or cannibalism happen late in life. Munch et al. (2005) and Jennings (2007) have independently developed a simple theoretic understanding of why density-dependent competition should happen late. Imagine a cohort with biomass $B_{\text{cohort}}(w)$ being spawned at a specific time. The individuals in the cohort will compete for food with other individuals from their own population (*intra*specific competition) or with individuals of similar size from other populations in the entire community (*inter*specific competition). If the resource is described by a Sheldon community spectrum, as in the consumer-resource model, the biomass of food in the community can to a good approximation be considered independent of body size (see p. 33). This means that all individuals have access to the same amount of food, B_{prey}, irrespective of their body size. Even though the amount of food is the same, the type of food is different—small individuals eat small food and large eat large food. Consequently, individuals compete for food only with individuals of similar size in the community. The degree of intraspecific competition versus interspecific competition—the degree to which individuals in the cohort compete among themselves rather than with other species—is the ratio between the biomass in the cohort and the biomass in the community: $B_{\text{cohort}}(w)/B_{\text{prey}}$. We saw in eq. 4.32 that the biomass of a cohort increases with size as $B_{\text{cohort}} \propto w^{1-a}$, where a is the physiological mortality and B_{prey} is independent of w. The degree of intraspecific competition, therefore, increases with size $\propto w^{1-a}$. Since a is typically < 1 (section 4.4), the exponent of the competition is positive and the degree of intraspecific competition increases with body size. The increase goes on until around the size at maturation, where the cohort biomass begins to decrease (fig. 4.3). The degree of intraspecific competition therefore reaches a maximum among adults and not among juveniles, as is clearly seen in fig. 10.4.

The argument developed here predicts that the brunt of density-dependent regulation occurs among adults. Competition among adults is commonly observed in populations of fish in small lakes, such as roaches, perch, or brown trout, where it leads to stunted growth (Burrough and Kennedy, 1979; Ylikarjula et al.,

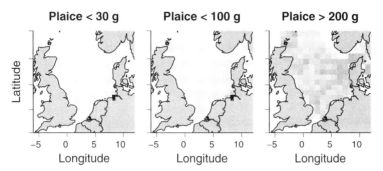

FIGURE 10.5. The spatial spreading of cohorts of North Sea plaice as they grow in size. Data from Andersen et al. (2017).

1999): the individuals are just able to reach maturation and then growth stops entirely. This type of density dependence among adults is clearly ill-represented by a stock-recruitment relationship that assumes early density dependence, with the implication that growth and mortality is independent of density. In marine environments, density-dependent changes in adult growth are observed in some cases (Lorenzen and Enberg, 2002; Zimmermann et al., 2018); however, the observed changes in growth are too small to lead to the extreme stunted growth observed in lakes. What is it in the marine environments that apparently defies the argument just developed?

The argument developed earlier about competition with other individuals in the community is very robust and hard to refute. It has disturbed me for many years because it implies that the use of stock-recruitment relationships is in serious error. However, the signatures of strong late density dependent regulation, such as a stunted adult growth, are rarely observed. The problem with the argument is a reliance on spatial homogenous population dynamics. The argument implicitly assumes that the cohort fills the entire habitat of the population as soon as it is spawned and thereby competes with the community within the entire habitat. This is a reasonable assumption for a small lake where newly hatched larvae quickly fill the littoral zone of the lake. In the ocean, or even in a semi-enclosed sea, the spatial element cannot be ignored. Eggs are typically spawned in one place and once larvae hatch they disperse, first passively through turbulent diffusion and later actively by horizontal movement; see fig. 10.5 for an example. I will now expand the argument about density dependence to consider a cohort spawned not only at a specific time but also at a specific place. For simplicity, we can describe the fraction of the total habitat that a cohort occupies as $\xi(w) \propto w^s$—that is, it increases with the size of individuals in the cohort with a rate determined by the exponent $s > 0$. As before, individuals will compete with other individuals but now only with those within the area that the cohort occupies $\xi(w)$. The competition will therefore be $B_{\text{cohort}}(w)/(B_{\text{prey}}\xi(w))$. Using again $B_{\text{cohort}}(w)/B_{\text{prey}} \propto w^{1-a}$, we get the competition to be $\propto w^{1-a-s}$. The scaling exponent is now reduced because individuals encounter more and more prey. If $s > 1 - a$, then the exponent of competition becomes negative, and the intraspecific competition will be most intense among the smallest juvenile individuals and not, as before, among adults. Therefore, if the dispersion of individuals is sufficiently fast, density-dependent competition will be most intense among juveniles. Dispersion can of course only continue until the individuals fill the entire habitat, at which point the spread of the population stops, and late-life density dependence might occur (Andersen et al., 2017). Accounting for the spatial aspect of density dependence not only provides an explanation for the observations of strong density-dependent processes in early life (Myers and Cadigan, 1993) but also explains why density-dependent processes will happen late in life in smaller habitats where the cohort immediately fills the entire habitat.

The spatial argument developed here is fairly simple, and mostly relevant for pelagic species spending all their life in the pelagic zone. A simpler argument to explain density-dependent regulation early in life is to consider limitation by juvenile habitat. Many species—in particular, demersal species such as a cod or plaice—settle in a demersal habitat after the metamorphosis at the end of their larvae phase. The competition for suitable habitats is often thought to be a strong density-dependent effect, and, at least for plaice in the North Sea, has been shown to determine recruitment (van der Veer, 1986). In any case, the preceding argument provides a theoretical justification for the use of stock-recruitment relationships that is internally consistent within the model framework itself, as well as with the process of juvenile habitat limitation. It also shows that the places where we mostly expect the stock-recruitment description to be in error are in small lakes and for large-bodied species.

10.4   FISHING ON A STOCK WITH EMERGENT DENSITY DEPENDENCE

Density dependence occurring late in life challenges one of the cornerstones of fisheries management—namely, the regulation of the minimum legal size of fish caught. Minimum mesh size regulation is born out of the theoretical insights from classic fish stock modeling, where yield is maximized by fishing adults. This theory was throughly analyzed in section 5.4 and fig. 5.11. Minimum landings sizes seem intuitively correct—making sure that fish are allowed to spawn at least once before they are caught will secure the stocks' reproductive output. Today, the idea of minimum size regulations are so entrenched in fisheries management that it has been forgotten where it comes from, and questioning it, however feebly, is considered sacrilegious. In 2013, the author of one of the standard fisheries textbooks, Carl Walters, was quoted as saying that, "Only 'a shrinking minority of fools' think that increasing fishing pressure on juveniles is smart or sustainable" (Borrell, 2013). Counterexamples from other practices where yield is maximised by harvesting juveniles are easy to find. If you sow carrots, you will lay many more seeds than you expect to have plants in the end. As the small seedlings grow, they begin to shade one another and their roots compete for nutrients. You can minimize the wasteful competition between the small plants by thinning. In other words, the production is maximized by harvesting both juvenile and adult plants. The production practice of maximizing yield by harvesting juveniles is also widely practised in forestry. Because of such considerations, the discussions back in the day about whether or not to fish juveniles were quite emotional. Sidney Holt describes it such (Holt, 2006):

> In, I think, 1948, at a Plenary Session of the Permanent Commission for Northeast Atlantic Fisheries, in London, the Commissioner for France—a biologist, M. Furnestin—waved two frying pans, a larger one and a smaller one. "This" he said (if my memory serves me correctly), shaking the smaller pan, "is a French pan; French housewives like to sauté small soles. And this—waving the bigger one—is an English (he didn't even say British) pan in which your housewives like to fry plaice, and bigger soles if they can get them."

The description beautifully reveals how rivalries between nations, culinary cultures, and scientific fields—Sidney Holt is a mathematician—all played a role in shaping fisheries management practice. In this section, I will run the risk of being in cahoots with foolish Francophile biologists by redoing the analysis of which size of fish we should fish to maximize the yield. The analysis is the same as in section 5.4, just now considering different types of density-dependent control.

Fig. 10.6 shows the yield from a stock that is exposed to trawl selectivity with a 50 percent selectivity at a fraction η_F of its asymptotic size (see fig. 5.2a for the trawl selectivity curve). With early density dependence ($R_{max} \ll \tilde{R}_{max}$), yield is maximized by fishing only adults. When density dependence happens late, through competition and cannibalism, the yield is maximized when also juveniles are caught. The effect is most evident when competition is the dominant effect. It is important to note that the maximum also becomes less pronounced. It does not matter much which size selectivity is used; the maximum yield will be more or less the same.

In fig. 10.7a, I mix early and late density dependence by varying R_{max}. Remember that small values of R_{max} lead to early density dependence (the classic situation), while large values leads to emergent late density dependence. Mixing in a bit of late density dependence does not change the classic result much.

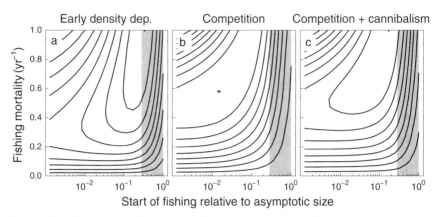

FIGURE 10.6. Yield from fishing as a function of the selectivity and the fishing mortality for three types of density dependence: (a) only stock-recruitment relation (compare with fig. 5.2); (b) resource competition; (c) resource competition and cannibalism. Asymptotic size $W_\infty = 1$ kg; the gray area indicates mature fish.

Only when density dependence is sufficiently strong, and only when cannibalism is absent, is yield maximized by fishing juveniles. The picture changes a bit when we look to stocks with different asymptotic sizes (fig. 10.7b). For small species, the classic results seem to hold regardless of the type of density dependence. For large species, the effects of late density dependence become more pronounced, also when cannibalism is present. For example, for a cod stock with an asymptotic weight around 20 kg with cannibalism and no early density dependence, yield is maximized by a trawl that targets sizes of 1 kg and larger.

Evidently, there is theoretical support for the notion that yield may be maximized by increasing the fishing pressure on juveniles. Fishing juveniles maximizes yield in populations with only late density dependence if the asymptotic size is larger than 200 g (2 kg if cannibalism is present, which it often is; see Fox, 1975). If early density is also present, then the asymptotic sizes where this happens are even larger. Further, the maximum is quite wide, so not fishing with exactly the gear that maximizes yield does not compromise yield much.

10.5 SUMMARY

This chapter was framed by two questions about density dependence. The first was whether it matters which type of density dependence operates in a model of a fish population. The answer is: absolutely! The size at which density dependence acts changes the demography. Increased mortality—for example, through

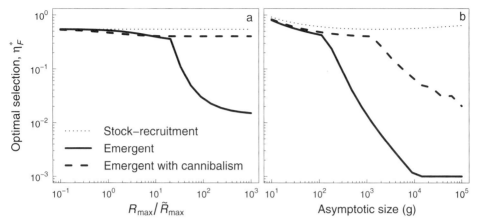

FIGURE 10.7. The fisheries selectivity that maximizes yield for situations with and without cannibalism (dashed and solid lines) compared to the optimal selectivity with a classic demographic model (dotted lines; see the middle column in fig. 5.4). (a) Optimal selectivity as a function of R_{max}; to the left density dependence is determined only by the stock-recruitment relation, and to the right density dependence is determined only by resource competition and cannibalism. Asymptotic size $W_\infty = 1$ kg. (b) Optimal selectivity as a function of asymptotic size.

cannibalism, reduces the abundance in the size range where the density dependence acts. Decreased growth will tend to increase the abundance in the affected size range because the slower growth rates lead to pile-up of individuals. The type of density dependence clearly also has potential repercussions for the minimum size regulations used in fisheries. If the density dependence acting late in life is sufficiently strong, then yield is maximized by fishing juveniles. However, I also showed that for most type of stocks the prediction from classic models, which assume only early density dependence, are probably sufficient. Only in cases of large-bodied stocks with strong density-dependent reductions in growth will the late density dependence affect minimum size regulations.

The second question was: what is the balance between the three types of density dependence in fish population in situ? This question cannot be answered in general; however, the simple theory provided some guidelines related to the size of the habitat and the asymptotic size of the stock. Stocks confined in a small habitat (small relative to their asymptotic size) are candidates for density-dependent regulation late in life. Further, large species are also expected to have at least some density-dependent regulation late in life. Demersal species, who settle in a benthic habitat around metamorphosis, are often limited by the size of the suitable juvenile habitat. This will induce some early life density dependence. However, they may experience additional late density dependence as well—in particular, the larger species. These are general considerations, but there will be big variations between stocks. A key aspect is the shape of the resource spectrum. I have assumed a nice, clean Sheldon spectrum. In reality, the size distribution of the resource will be bumpy, and this may introduce bottlenecks that lead to competition and reduced growth of particular sizes.

In terms of application to fisheries science, the model I developed and analyzed in chapters 4 and 5 is fundamentally similar to the well-known age-based model, just reformulated in terms of size instead of age. That reformulation gave us additional insights into model behavior and dynamics, but in terms of fundamental assumptions the model is old school. Within population ecology, such a model would be considered of limited utility because it does not consider the interaction between the consumer (the fish population) and the resource (the zooplankton and prey fish). The size-based consumer-resource based model developed here shows how classic demographic principles from fisheries science can indeed be merged with the traditional consumer-resource picture from population ecology. The model is more complicated than classic unstructured consumer-resource or predator-prey type of models because it has to resolve the ontogenetic trophic niche shifts of fish, but it is based on very few basic assumptions. Such a model is therefore the key to unite the thinking and methodology between fisheries science and population ecology. In the next chapter, I will also show how the model is the central building block in developing a model of the entire fish community suitable for implementing ecosystem approaches to fisheries management.

Whether the size-based consumer-resource model is the future framework for single-stock fisheries management requires that it can be operationalized in practical management settings. To do so, we must be able to obtain credible descriptions of the resource landscape for a fish stock (the resource size spectrum), and how it changes with time due to environmental changes and to competition with other species. This is a major undertaking, and time will tell whether this model can be operationalized in the twenty-first century to the same degree that the Beverton-Holt model was operationalized in the twentieth century.

CHAPTER ELEVEN

Trait Structure of the Fish Community

The beginning of this book, chapter 2, described the community size spectrum. The community size spectrum was predicted to depend on the exponents of metabolism n and clearance rate q as $N_c \propto w^{n-2-q}$. The community size spectrum describes the size distribution of organisms of different sizes, but tells little about which species they belong to. The aim of this chapter is to combine community spectrum with the size spectra of the populations in the community to develop a more detailed theory of the structure of the fish community.

While working with single stocks, I emphasized asymptotic size W_∞ as a master trait. I argued that by just knowing W_∞, it was possible to deduce the entire set of parameters for the stock on the basis of life-history parameters that do not vary systematically between fish stocks. While obviously inaccurate for a specific stock, where the parameters may deviate from the average life history, the use of W_∞ as a master trait paved the way for sweeping statements about the demography and impacts of fishing broadly across all species. In this chapter, I will use W_∞ as a trait-axis to establish a *trait distribution*. Specifically, I will calculate the abundance (or biomass) of all species in a community as a function of their asymptotic size. How that works concretely will be clear shortly.

Next, I will develop a dynamic model of the fish community that will be used for ecosystem-level impact assessment of fishing in the next chapter. The core of the model is the consumer-resource model from chapter 10. Linking a string of consumer models with different asymptotic sizes creates a description of the entire community. Only a few additional steps are needed to glue these models together into a full dynamic community model of the fish community.

The chapter is organized in three main sections. In the first section, I develop a purely analytical theory of the asymptotic size trait distribution in a fish community. The theory is based upon the Sheldon community spectrum developed in chapter 2, and I will use the new theory to formulate an "extended" Sheldon conjecture. The analytic theory describes only a steady-state solution, which is of limited use for impact assessments of fishing; that requires a dynamic trait-based size spectrum model, which is developed in section 11.2. Last, I show how the trait-based model can be extended to model specific stocks embedded in a food web.

11.1 STRUCTURE OF AN UNFISHED COMMUNITY

What is the structure of a fish community? That depends on what one reads into the word *structure*. Chapter 2 explored the size structure of an ecosystem through the community size spectrum—that is, the abundance (or biomass) of all individuals as a function of their size only (section 2.4). The community size spectrum is just one aspect of community structure. Another obvious aspect is the populations: which populations make up the community and what are their abundances? At least we would like the structure to explain something about the population structure—if I pick a random 10 g fish, is it then an adult forage fish or a juvenile top predator? Digging further down into the structure, the populations themselves have a size structure, as calculated in chapter 4. The size structure of populations is different from that of the community—the community abundance size spectrum is a power law with exponent around -2 (section 2.4), while the populations have an exponent around -1.1 and are truncated at the asymptotic size (eq. 4.27). Nevertheless, the two spectra are related: the community is the sum of all the population spectra. We can therefore derive the community size spectrum, $N_c(w)$, by integrating over the size spectra of all populations with an asymptotic size larger than w

$$N_c(w) = \int_w^\infty N(w, W_\infty)\, \mathrm{d}W_\infty. \tag{11.1}$$

While the preceding relation appears trivial, it harbors an important conceptual extension of the size spectrum concept: the population size spectrum is not only a function of mass w, it is also a function of asymptotic mass W_∞. The distribution $N(w, W_\infty)$ is a combination of a *size distribution* along the w axis and a *trait distribution* along the W_∞ axis. Being a trait distribution means that $N(w, W_\infty)\Delta W_\infty$ represents the abundance of all populations with asymptotic sizes in the range $[W_\infty : W_\infty + \Delta W_\infty]$. Similarly, the number of individuals in those species within a range of body sizes $[w : w + \Delta w]$ is $N(w, W_\infty)\Delta W_\infty \Delta w$. The extension from a size distribution $N_c(w)$ to a trait distribution $N(w, W_\infty)$ entails a change of dimensions. Where the abundance size spectrum had dimensions of number per body mass per volume, the size-trait distribution has dimensions of number per body mass per volume per asymptotic size. The size and trait distribution $N(w, W_\infty)$ provides a simple description of both size and population structure of the community.

We can use the relation between the community spectrum and the population spectra in eq. 11.1 to calculate the abundance of populations within a range of asymptotic sizes. On the left-hand side, we can insert the community spectrum found in section 2.4: $N_c(w) = \kappa_c w^{-2-q+n}$, where $q = 0.8$ is the exponent of the clearance rate and $n = 3/4$ is the metabolic exponent. On the right-hand side, we can use the population size spectrum from chapter 4. There, we calculated the size spectrum of juveniles in a population as $N(w) = C w^{-n-a}$, where $a \approx 0.34$ is the physiological mortality, and C is an unknown constant that determines the overall abundance of the population (case I in box 4.3). Assuming that C is a power-law function of W_∞, $C(W_\infty) = \kappa W_\infty^d$, we can insert it in eq. 11.1 to solve for the exponent d

$$N_c(w) = \int_w^\infty C(W_\infty) w^{-n-a} \, \mathrm{d}W_\infty \Leftrightarrow \tag{11.2}$$

$$\kappa_c w^{-2-q+n} = -\frac{\kappa}{1+d} w^{-n-a+1+d} \quad \text{for} \quad 1+d < 0 \Leftrightarrow \tag{11.3}$$

$$d = 2n - 3 - q + a \approx -1.96 \quad \text{and} \tag{11.4}$$

$$\kappa = (2 + q - 2n - a)\kappa_c \approx 0.88\kappa_c. \tag{11.5}$$

In box 11.1, the calculation is extended to the more complicated solution of the size spectrum found using the von Bertalanffy growth equation. The result for the exponent d is the same, but the relations between κ and κ_c are slightly different.

We now have a general solution for the size and trait distribution of a marine ecosystem

$$N(w, W_\infty) \propto W_\infty^{2n-3-q+a} w^{-n-a} F(w/W_\infty), \tag{11.6}$$

where the last function $F(w/W_\infty)$ describes the shape of the population size spectrum; see the square brackets in eq. 11.7 for an example. What does that solution tell us about community structure? It shows that the abundance scales with asymptotic size, W_∞, with exponent $2n - 3 - q + a \approx -2$. In logarithmically spaced asymptotic size groups—that is, with the width of the group proportional to W_∞, the abundance or biomass then scales roughly as W_∞^{-1} (see box 2.1). This means that for a given body size, there are fewer fish with a large asymptotic size than with a small asymptotic size. Or, in plain words, there is a larger abundance of 10 g herring-like fish than 10 g cod-like fish.

The pattern predicted by eq. 11.6 is borne out in real fish communities. Fig. 11.1 shows two representations of the size structure of the North Sea fish community compared to the analytical solution. The first representation is straightforward: it is simply the biomass spectra of the 9 dominant species varying in asymptotic size from 53 g (sand eel) to 17 kg (cod). Clearly, smaller species are more abundant than larger species when compared at the same body size. There are, of course, also examples of species with very low biomass—sole has the lowest spectrum

BOX 11.1

ANALYTICAL SOLUTION OF ASYMPTOTIC SIZE TRAIT DISTRIBUTION

Box 4.3 p. 65 derived an analytical solution to the population size spectrum (eq. 4.14), which can be written as

$$N(w, W_\infty) = \kappa W_\infty^d w^{-n-a} \left[1 - \left(\frac{w}{W_\infty} \right)^{1-n} \right]^{\frac{a}{1-n}-1}, \quad (11.7)$$

where some of the constant terms have been absorbed into the constant κ and I have made the *ansatz* $R = W_\infty^d$. It is the exponent d that we aim to derive. Inserting the community spectrum $N_c = \kappa_c w^{-2-q+n}$ and eq. 11.7 into eq. 11.1 gives

$$\kappa_c w^{-2-q+n} = \int_w^\infty \kappa W_\infty^d w^{-n-a} \left[1 - \left(\frac{w}{W_\infty} \right)^{1-n} \right]^{\frac{a}{1-n}-1} dW_\infty. \quad (11.8)$$

Changing integration variable to $x = w/W_\infty$ gives

$$\kappa_c = \kappa \int_0^1 x^{-d-2} w^{d-2n-a+q+3} (1 - x^{1-n})^{\frac{a}{1-n}-1} dx. \quad (11.9)$$

As w enters in only one term in the integral, that term has to be constant. This only occurs when $d = 2n + a - q - 3$. The integral can then be solved to give

$$\kappa = \kappa_c a \frac{\Gamma \left(\frac{1-a+d}{n-1} \right)}{\Gamma \left(1 + \frac{a}{1-n} \right) \Gamma \left(\frac{1+d}{n-1} \right)} \quad (11.10)$$

for $0 < n < 1$, $a > 0$ and $d < 1$, all of which are satisfied. Standard values of the parameters ($n = 0.75$, $q = 0.8$, and $a = 0.42$) give $\kappa \approx 2.65\kappa_c$—about a factor of 2 higher than the simple calculation in eq. 11.5.

but middle-range asymptotic size.[1] This prediction is similar to the one made by the theory; however, the theory makes no predictions about population biomass, it only predicts the biomass of trait groups—that is, of all species with comparable asymptotic sizes. In the middle panel, the populations are lumped together in trait groups with asymptotic sizes from 10 to 100 g (sand eel and Norway pout),

---

[1] The data I use here are not direct observations, but output of a complex statistical assessment model, the SMS model (Lewy and Vinther, 2004). Had the SMS model resolved all species in the North Sea, there would be many more species with small biomass. The SMS model is designed for fisheries management, and the model resolves only the species of commercial interest. As the fishery in the North Sea is fully developed, the commercial species represent most of the biomass. A relatively rare species like sole is included in the SMS model only because of its high commercial value.

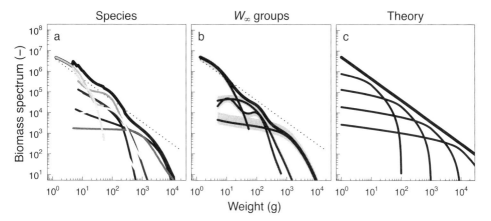

FIGURE 11.1. Three representations of size spectra in the North Sea: population spectra (a), asymptotic size group spectra (b), and the theoretical solution from box 11.1 (c). The data in panels a and b are output from the SMS multispecies statistical model (Lewy and Vinther, 2004) that is routinely used to estimate species interactions for the management of North Sea fish stocks. The model estimates the main commercial species (W_∞ in parentheses): sand eel (53 g), Norway pout (56 g), herring (260 g), sole (697 g), whiting (1,015 g), plaice (1,025 g), haddock (4,701 g), saithe (11.2 kg), and cod (16.7 kg). The lines for each spectrum in (a) are from the year 1990. Panel b groups species by asymptotic sizes in ranges 10 to 100 g, 100 to 1000 g, 1 to 10 kg, and 10 to 100 kg; the gray areas represents the variability in the period 1975–2008. The spectra and asymptotic size groups are fitted to the SMS model output with the generalized additive model: $\log N_i(\log w, t) = \text{constant} + s_{i,1}(\log w) + s_{i,2}(t)$, where $s_{i,j}$ are splines for species i and t is time.

100 to 1000 g (sole and herring), 1 to 10 kg (whiting, haddock, and plaice) and larger than 10 kg (saithe and cod). The qualitative features of the trait distribution in panel b are similar to the theoretical prediction in the right panel. Even the quantitative prediction of a reduction in biomass by a factor of 10 for each factor of 10 asymptotic size group is seen in the data. This correspondence between data and theory is remarkable considering that the theory is based on just three very general assumptions: the metabolic scaling of consumption with exponent $n \approx 0.75$, the scaling of clearance rate with exponent $q \approx 0.8$, and the rule that larger individuals eat smaller individuals. There are also deviations between data and theory. The most evident is that the community spectrum of the North Sea community is steeper than the theoretical prediction. This deviation is a result of the heavy fishing pressure. The next chapter is devoted to describing such effects of fishing, but the following text is devoted to a deeper exploration of the analytical solution.

While describing the community size spectrum in chapter 2, we found the exponent $-2 - q + n$. This exponent corresponds to the biomass in logarithmically distributed size groups having exponent $-q + n \approx -0.05$—that is, being almost independent of body size. This is the celebrated Sheldon spectrum (p. 15 and fig. 2.2).

With our newly found structure of the community eq. 11.6, we can go even further. By integration of eq. 11.6, we can calculate the total biomass of all life histories with asymptotic sizes in the range W_∞ to kW_∞, where k is an arbitrary constant:

$$B_W(W_\infty) = \int_{W_\infty}^{kW_\infty} \int_0^{W_\infty} N(w, W_\infty) w \, dw \, dW_\infty \propto W_\infty^{n-q} \approx W_\infty^{-0.05}. \quad (11.11)$$

The result is identical to the Sheldon spectrum only with individual size w replaced by asymptotic size W_∞. With that result, we can formulate an extension to the Sheldon conjecture, as follows:

The total biomass of species within logarithmically spaced asymptotic size groups is approximately constant.

The extended Sheldon conjecture means that the total biomass of all species with asymptotic sizes between, say, 10 and 100 g is approximately the same as that of species with asymptotic sizes in the range 1 to 10 kg.

No empirical test exists of the extended Sheldon conjecture. The closest is an analysis of trawl survey data from the North Sea by Daan et al. (2005). They calculated the abundance of all individuals in asymptotic size groups. Unfortunately, they did not calculate the biomass, which is problematic, as I will show. The abundance of a population is dominated by the smallest individuals in the population—there are more larvae than adults. This statement emerges by integrating the population size spectrum $N(w) \propto w^{-n-a}$ over a size range from w to kw. This calculation gives the abundance in a logarithmic size group as $\propto w^{1-n-a} \approx w^{-0.2}$. The abundance is a decreasing function of size and is therefore dominated by the smallest individuals. The biomass in the same size range is $\propto w^{2-n-a} \approx w^{0.8}$—that is, increasing. The biomass is therefore dominated by the adults. The estimation of the juvenile abundance from the trawl survey is expected to be more uncertain than the estimation of adults, so the calculation of abundance gives a less robust result. Anyway, we can still use the theory to calculate the abundance within logarithmically spaced asymptotic size groups up to a constant factor of proportionality as

$$N_W(W_\infty) = \int_{W_\infty}^{kW_\infty} \int_{w_{min}}^{W_\infty} N(w, W_\infty) w \, dw \, dW_\infty \quad (11.12)$$

$$\propto \frac{1 - w_{min}^{1-n-a}}{n - 1 - q} W_\infty^{n-1-q}. \quad (11.13)$$

Note that I specified the lower limit of the inner integral explicitly as w_{min}. This is necessary because, as argued earlier, the abundance increases as the lower size decreases. Without an explicit lower limit, the integral would diverge to ∞. When calculating the biomass in eq. 11.11, the lower limit becomes insignificant because

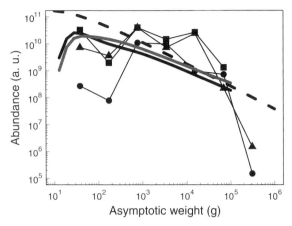

FIGURE 11.2. The total abundance of all species grouped according to asymptotic size. The points connected with thin lines are from three different bottom trawl surveys in the North Sea (Daan et al., 2005). The three thick lines are calculations: based on theory (as in eq. 11.11; dashed line) and from the trait-based model with and without fishing (gray and black lines; the model is introduced in section 11.2). These calculations count only individuals larger than 10 cm, which is the smallest size retained in the trawl surveys. The lines are multiplied by arbitrary constants to account for the unknown catchability of the trawl surveys. Therefore, only the shapes of the curves can be compared with the data points, not the absolute level.

the biomass is concentrated among the adult population. The comparison with the trawl surveys is not particularly convincing (fig. 11.2). The observations show a more dome-like structure with smaller abundance of the smallest and largest species than the theoretical prediction. The lower abundance of smaller species could well be due to the sampling artifacts discussed earlier. The lower abundance of larger species simply reflects that absence of very large bodied species in the North Sea due to fishing. Therefore, with the inherent uncertainties due to the calculation of the abundances and not biomass, it is difficult to use the extant data to reliably confirm or falsify the extended Sheldon conjecture.

11.2   DYNAMIC COMMUNITY MODEL

The analytical solutions of the community size spectrum (chapter 2), the community structure (eq. 11.5), and the extended Sheldon conjecture (eq. 11.11) are aesthetically pleasing mathematical derivations from just a few basic assumptions: metabolic scaling of consumption, scaling of clearance rate, and the rule that bigger fish eat smaller fish. Their simplicity is also their limitation: they are not useful for making quantitative predictions about the impact of fishing. Nevertheless, the analytical solutions provide null hypotheses for the structure of

BOX 11.2

LEVEL OF RECRUITMENT

We can determine the maximum recruitment by combining the relation between recruitment flux from the boundary condition eq. 4.18: $g(w_R)N(w_R) = R$ with eq. 11.6 and the simple expression for juvenile growth $g(w_R) = Aw_R^n$ to get: $R \propto AW_\infty^{2n-q-3+a}w_R^{-a}$, with w_R being the size at recruitment. That relation gives the recruitment of a single population. The entire group will scale with the overall magnitude of the resource spectrum, κ_{res}, and with the range of asymptotic sizes that the group represent $\Delta W_{\infty.i}$

$$R_{max.i} = K_{Rmax}\kappa_{res}AW_{\infty.i}^c w_R^{-a}\Delta W_{\infty.i}. \qquad (11.14)$$

K_{Rmax} is a nondimensional factor that determines the strength of early-life density dependence imposed by the stock-recruitment relation. We expect it to have a value around 1: a higher value means that the stock-recruitment imposes only weak density-dependent control, while smaller values imply higher imposed control. How much external density dependence does it take to ensure coexistence between asymptotic size groups, and how small should we make K_{Rmax}? Fig. 11.3 shows how at high values ($K_{Rmax} > 0.5$) some asymptotic sizes are excluded—notably, the medium sizes with asymptotic sizes around 100 g—while smaller values lead to coexistence. In the following, I use the largest value that ensures coexistence: $K_{Rmax} = 0.25$.

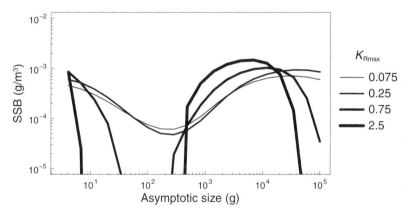

FIGURE 11.3. Spawning stock biomass as a function of asymptotic size for various values of the coefficient K_{Rmax}: at low values, density dependence is determined mainly by the stock-recruitment relation; at high values, density dependence occurs mainly by the internal processes of competition and predation in the model, which leads to exclusion of species with asymptotic sizes around 100 g.

the marine ecosystem that serve as benchmarks to compare with data or more complex models.

Making predictions about fishing requires a dynamical model. The model will be a trait-based model—that is, it will model the trait distribution $N(w, W_\infty)$, and not the specific populations. The W_∞-axis of the distribution is discretized into W_∞-groups, just as in fig. 11.1, and each W_∞-group is represented by the consumer-resource model from section 10.1. The resource, which was earlier described by the community spectrum, is replaced by the sum over all the W_∞-groups. In this way, most of the resource is replaced by the solution of the model itself. The resource now represents only the smaller nonfish part of the community. The resource is still represented by a spectrum, but the spectrum is cut short at a size $w_{cut} = 2$ g. Other minor differences to the consumer-resource model are a slight adjustment of the clearance rate, a background mortality that scales with W_∞, and a specification of how R_{max} scales with W_∞ that I will describe in the following (summarized in appendix C).

Recruitment

The population model in chapter 10 included a description of density dependence though a stock-recruitment function. In the community model, density-dependent effects of food competition and cannibalism are explicitly represented, and a stock-recruitment function should not be needed. Unfortunately, things are not that simple. There are two complications at play: first, as discussed in section 10.3, population dynamics with consumption and mortality described by "metabolic" scalings ($\propto w^n$ and w^{n-1}, respectively) implies that density dependence occurs mostly late in life, around maturation size. Within the model, late density dependence will manifest itself as stunted growth, which is not common among marine fish populations. Consequently, some density dependence should occur early in life. Second, if the model is run without externally imposed density dependence, there will be competitive exclusion between the species. Typically, only a small and a large species will persist (Hartvig and Andersen, 2013). The reason is that there is insufficient niche differentiation between the species to allow for coexistence. Allowing for coexistence requires that species are differentiated by more than just their asymptotic size. These two effects means that some additional density dependence is needed, which should preferably act early in life. The standard way to acheive that is through a stock-recruitment relation, just as in chapters 4 and 10.

The stock-recruitment relation specifies the recruitment of the ith asymptotic size group as a function of egg production as (see also fig. 4.4 and eq. 4.36):

$$R_i = R_{max.i} \frac{R_{egg.i}}{R_{egg.i} + R_{max.i}}. \tag{11.15}$$

When working out the demography of a single stock in part II, we could ignore the maximum recruitment and just specify all population-level metrics—biomass, size spectrum, yield, and so on—relative to the maximum recruitment. Here, the absolute biomasses of the asymptotic size groups matter, and we need to specify maximum recruitment for each group—that is, how R_{max} varies as a function of asymptotic size. That is a tough nut to crack! There is no biological knowledge to guide us about how the total recruitment of all species with similar asymptotic sizes should vary with asymptotic size. I will adopt a pragmatic solution that ensures coexistence between all asymptotic size groups. The solution is inspired by the analytic solution in section 11.1 and the observation of how biomass varies with asymptotic size (fig. 11.1b): I assume that the maximum recruitment scales with asymptotic size, as does abundance in the analytical solution of the community in eq. 11.6 and fig. 11.1c: $R_{max.i} \propto W_\infty^{2n-q-3+a}$. In box 11.2, I describe how to approximate the constant of proportionality, which sets the level of externally imposed density dependence.

The need for a mechanism to stabilize coexistence is not unique to the trait-based size spectrum model. Almost all food-web-type models need such a mechanism. One of three mechanisms is commonly invoked: random food-web matrices, carrying capacities, or switching. Each mechanism has with its own pros and cons.

A random food-web matrix means that interactions between individuals are characterized by a species-to-species interaction coefficient in addition to the size-based interaction. Such interaction matrices can be purely random (Hartvig et al., 2011) or be constrained by some structure that better represents the structure of natural food webs (Pimm et al., 1991; Petchey et al., 2008; Hartvig, 2011; Rossberg, 2013; Zhang et al., 2014). Using a species-based trait (the interaction) departs from the idea of the pure trait-based model, which is exactly to avoid describing population dynamics at the species level. Further, analysis of the results requires averaging over many realizations of the random food-web matricies, which is cumbersome. Using a species-level random food-web matrix is therefore an unpleasant option.

Some species-based food-web models are designed to represent real food webs with specific species from a particular region. Here, again, the food-web model is unlikely to ensure coexistence between all species without a stabilizing mechanism. That can be a carrying capacity (De Ruiter et al., 1995), interference competition for each species (Zhang et al., 2015), or a predator-dependent functional response (Walters et al., 2000).[2] As these mechanisms correspond to

---

[2] The mechanism used in Ecosim is presented as "foraging arena theory." In practice it is a very close cousin of ratio-dependence (Barraquand, 2014).

having a carrying capacity for each species, they are essentially similar to the stock-recruitment relationship that I have resorted to.

The mechanism with the best grounding in ecological theory is *prey switching* (used by Maury and Poggiale, 2013). Prey switching means that a predator preferentially targets more abundant species over less abundant species (Murdoch, 1969; Murdoch et al., 1975). Consequently, the more abundant species experience higher predation mortality than the less abundant ones. In this manner, population abundances will be equilibrated between species such that, eventually, 100 g cod-like species (large asymptotic size) will be as abundant as a 100 g herring-like species (small asymptotic size). This, however, differs from the observations from the North Sea (see fig. 11.1), where a 100 g cod-like species is less abundant than a 100 g herring-like species. Switching will therefore lead to a different community structure than observed (see figure 4 in Maury and Poggiale, 2013), which is why I have not used switching.

The conclusion that emerges is that there are several ways to represent the mechanism that stabilizes coexistence in food-web models. Currently, we have no way to determine which is the more ecologically correct way. Probably all effects—carrying capacity early in life, interference, and switching—operate simultaneously to varying degrees in different populations and ecosystems. The choice of using a stock-recruitment relationship is therefore not the last word in this story. It has the advantages that it induces some density-dependent control early in life that we indeed observe among marine fish populations, and that the scaling of R_{max} has some theoretical support. The solution can be perceived as the "least bad" among many choices.

11.3   DYNAMIC COMMUNITY MODEL VERSUS ANALYTIC THEORY

The model is a trait-based model with asymptotic size being the trait. In principle, the asymptotic size trait distribution is continuous, but for the numerical solution the asymptotic axis has been discretized into asymptotic size groups. Does it matter how many asymptotic size groups one chooses? Not much, it turns out. Most examples given in this chapter use nine asymptotic size groups—that is sufficient to cover the variation along the entire range of asymptotic sizes used (from 4 g to 100 kg). Choosing more groups will lead to smoother results but no qualitative differences.

The structure of the solution is revealed by the size spectra, the feeding level, the mortality, and the eggs per recruit (R_0; see section 4.2), as shown in fig. 11.4. The solution is compared to the theoretical solution from section 11.1 with dashed lines. The results of the dynamic model undulate around the theoretic predictions

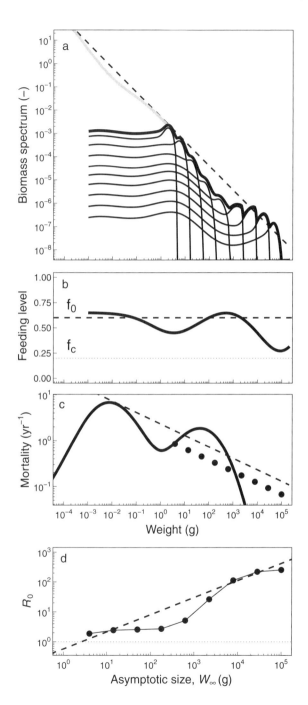

(dashed lines). This undulation is largely driven by the truncation of the spectrum at the largest body sizes. The largest individuals do not experience any predation mortality in the model, they only die due to a small background natural mortality that represents disease, senescence, and predation from marine mammals. I assume that the background mortality scales with asymptotic size (the black dots in fig. 11.4c):

$$\mu_b = \mu_0 W_\infty^{n-1}. \tag{11.16}$$

The level of the background mortality is controlled by the parameter μ_0, which is set such that the background mortality is lower than the theoretical predicted predation mortality. This means that the largest individuals in the model experience smaller losses than the rest of the fish community. They are therefore superabundant relative to the theoretic prediction (higher abundance by about a factor of 2). This superabundance has a number of consequences. First, the large individuals face more competition for their prey, resulting in a lower feeding level of individuals larger than about 10 kg. The ensuing reduced growth leads to an accumulation of the larger individuals, which further swells their numbers. Second, the superabundance of the largest individuals induces a higher predation pressure on their prey in the size range 10 to 100 g than the theoretic prediction (fig. 11.4c), such that there are fewer individuals in that size range than predicted by the theoretic solution. This, in turn, means that the 10 to 100 g fish induce a smaller predation pressure on their prey—mainly zooplankton from the resource—which are then depressed in abundance.

Is the preceding prediction of an unperturbed natural fish community correct? Is there superabundance (relative to the Sheldon spectrum) of the largest fish? Unfortunately, there are few unperturbed systems left where we can look at for guidance. One compelling example is provided by coral reef fish on remote Pacific islands, as analyzed by Stevenson et al. (2007). They showed that for islands undisturbed by fishing, up to half the fish biomass was of fish larger than 50 cm. This indicates that undisturbed size spectra are indeed "top heavy." This justifies the low value of the parameter μ_0 that controls the strength of the trophic cascade in the unfished system.

FIGURE 11.4. Results of the dynamic trait-based model compared to theoretical predictions (dashed lines). (a) Biomass spectra of the resource (gray), the community (thick line), and spectra of nine asymptotic size groups (thin lines). (b) Feeding level, with the dotted line indicating the critical feeding level where consumption is insufficient to meet basal metabolic demands. (c) Mortality from predation (thick line) and background mortality on each asymptotic size group (dots). (d) Eggs per recruit (R_0) measuring the degree of density-dependent regulation by the stock-recruitment relation. The dotted line at $R_0 = 1$ indicates the extinction point.

Another striking feature of the solution is the relatively low abundance of fish larvae—that is, individuals in the size range from 10^{-3} to 0.1 g. This means that fish larvae are unable to exert a significant predation pressure on microzooplankton. Or, in other words: fish larvae are rare in the ocean. This rarity makes density dependence less likely for the fish larvae. In the model, density-dependent effects appear at a size around 1 g. I discussed this in section 10.3 as a possible explanation for the early life density dependence.

The smallest size where food competition manifests itself within the model is around a few grams, at the size of small adult forage fish. The signature of food competition presents itself as a lowered feeding level in the range from 0.2 to 20 g. That period of depleted food is felt by species of all asymptotic sizes and the consequences are evident in the form of reduced egg production relative to the theoretical prediction (fig. 11.4d) (R_0 is less than it would have been if feeding level was constant at f_0). Within this range of asymptotic sizes, $R_0 \approx 3$. This means that the stock recruitment relation does not have much of an effect on these species (see fig. 4.4, where a value of $R_0 = 3$ means that the stock-recruitment relation deviates only slightly from the straight rising line at the origin)—most of the density dependence is taken care of by the competition for food. The stock-recruitment relationsship therefore mainly plays a role for fish species with a asymptotic sizes larger than around 1 kg.

11.4   SPECIES VERSUS TRAITS

The trait-based model does not resolve the actual species; it calculates the biomass of a range of species defined by asymptotic sizes. This is a problem for a management whose concerns are the specific stocks: how many cod (or tuna, or hake, and so on) are there, and how are they related to state of the herring stock (or anchovy, or sprat, and so on)? Further, the model assumes that species differ only in their asymptotic size. As discussed in section 3.4, species differ in other ways. A good example is rockfish of the family Sebastidae. Rockfish have a high value on the market, but they grow slowly, and some species are among the longest living fish in the oceans. Their relatively slow growth rate means that they have a lower consumption rate than species of similar asymptotic body size. Their impact on the community is therefore less than similar-sized species. However, of management importance, rockfish tolerate much less fishing mortality. Representing the difference in growth rate, and all the other differences between species, requires a more sophisticated model. The obvious extension is to increase the dimension of trait-space to also include a trait representing growth rate—for example, the maximum consumption rate h. This extension is yet to be made. Another extension is to move from a trait-based model to a food-web model, where each species is represented with its own size spectrum. I here sketch how the trait-based model can be turned into a species-based food-web model. However, for the impact assessment of fishing in the next chapter, I will use only the trait-based model.

Turning the trait-based model into a food-web model is conceptually straight-forward: we can simply interpret the trait-groups as species (Blanchard et al., 2014; Andersen et al., 2016; Jacobsen et al., 2017). Additional realism can be introduced by using species-specific parameter values instead of the average value used in the trait-based model, at least for the species where these parameters are known. The key parameters are the maximum recruitment, R_{max}, and the recruit-ment efficiency, ε_R. They represent all variation in processes not explicitly covered by the model and that determine the carrying capacity of the species: limitations in juvenile habitat, variation in recruitment success, and so on. Secondary in impor-tance are parameters related to growth—in particular, the maximum consumption rate, h. Next up is the predator-prey interaction between species on top of the size-based interaction, as mediated by, for example, differences in habitat over-lap. A pelagic species, for example, is unlikely to feed on a benthic species, and vice versa. Box 11.3 describes the practical aspects of setting up such a size-based food-web model.

BOX 11.3

CALIBRATION OF A FOOD-WEB MODEL

A species i in a food-web model is characterized by a set of parameters: recruitment: $R_{max.i}$ and $\varepsilon_{r.i}$; preference towards prey species j: θ_{ij}; growth: h_i and γ_i; and prey size preference: β_i, σ_i. Other parameters can be varied—for example, offspring size or size at maturation—but they have a modest influence on the model results.

The maximum recruitment, R_{max}, is the most important parameter for a species, as it determines its level of biomass. R_{max} can be estimated roughly by either using the maximum observed level of biomass or by calibration to observed catches (Blan-chard et al., 2014) or biomasses (Spence et al., 2015; Jacobsen et al., 2017). The second parameter of the stock-recruitment relation, ε_R, determines the recruitment success of the species. This parameter can be used to calibrate the fisheries reference points—a low efficiency means low resilience to fishing and vice versa (Jacobsen et al., 2017). If ε_R is not calibrated for each species, it should be checked that the fisheries reference points are in the right ballpark. Getting the reference points rea-sonably right is particularly important if the calibrated model will be used to explore fisheries scenarios.

(continued)

(Box 11.3 *continued*)

As we saw in fig. 3.3, there can be a large variation in growth rate between species with the same asymptotic size. In the dynamic model, growth rate is determined by the maximum consumption rate h. By combining the relation between observed von Bertalanffy growth constant K and A (eq. 3.10) with the relation between A and h (eq. 3.31), we get

$$h \approx \frac{3c^{1/4}\eta_m^{-1/12}}{\varepsilon_a(f_0 - f_c)} KL_\infty^{3/4} \approx 4.4KL_\infty^{3/4}, \tag{11.17}$$

where c is the constant that relates weight and length (p. 19). In the last approximation, K is measured in units of year^{-1}, L_∞ in cm, and h as g$^{1/4}$yr^{-1}. Note that the value of h also changes the clearance rate parameter γ (see table C.1). Differences in growth rates should also be reflected in a species' vulnerability to predation—fast-growing species are more vulnerable than slow-growing species (see the discussion of the physiological mortality on p. 78). Reduced vulnerability can be incorporated as an adjustment of the preference between predators i and prey j

$$\theta_{ij} = \frac{h_j}{\bar{h}}, \tag{11.18}$$

where $\bar{h}$ is the mean value of h_i of species in the food web.

With these changes, the two central equations of the model become

$$E_{e,i}(w) = V_i(w) \int_0^\infty \phi(w_p/w) \left(N_R(w) + \sum_j \theta_{ij} N_j(w_p) \right) w_p \mathrm{d} w_p, \tag{11.19}$$

which also influences the predation mortality

$$\mu_{p,i}(w_p) = \sum_j \theta_{ij} \int \phi(w_p/w)(1 - f_j(w))\gamma_j w^q N_j(w) \, \mathrm{d}w. \tag{11.20}$$

Examples of calibrated ecosystems can be found in Jacobsen et al. (2017).

Variations of this approach are currently being explored in the literature, from the simple calibration (Kolding et al., 2016; Jacobsen et al., 2017; Szuwalski et al., 2017), where only size-based interactions are used, over the more complex, with species-based interactions (Blanchard et al., 2014; Spence et al., 2015). Common to all of them is that they produce a food-web model with biomasses of specific named species. What can such a model be used for? One needs to be aware that despite a fancy statistical calibration method, the description of the population dynamics of each species is fairly crude. Using the model to predict the dynamics of a specific species should therefore be treated with caution. However, the model is still useful for general ecosystem-level impact assessments of proposed strategies for management objectives for a specific ecosystem. For example, evaluation of how indicators, such as the size spectrum exponent or the ratio of large to small fish (Greenstreet et al., 2010), change when fishing is increased; assessment of the consequences of F_{msy} management (Blanchard et al., 2014; Szuwalski et al., 2017); or assessment of the ecosystem-level efficiency of fisheries (Jacobsen et al., 2017).

11.5   SUMMARY

The development of a full dynamic community model is the crowning achievement of the size- and trait-based theory. The model connects everything developed so far into a single coherent model. Despite being a fairly complex model—it is conceptually simple. It is based on just two central assumptions: big fish eat smaller fish and the size-scaling of clearance rate. The earlier work in parts I and II also involved the metabolic assumption about consumption scaling as w^n. In the dynamic model, this assumption is relaxed somewhat, as consumption can vary with size and time. However, the metabolic assumption is not completely gone, because it still features in the maximum consumption rate hw^n.

The model has a small set of parameters. What is more important than the number of parameters is that the results are not sensitively dependent upon the value of the parameters. For example, the coefficient for maximum consumption h mainly scales time in the model—faster maximum consumption leads to faster growth rates and faster dynamics, but no big changes in the equilibrium situation. Similarly, the constant for the resource spectrum κ_{res0} mainly scales the overall abundance. The robustness of the model also indicates that the structure of natural fish communities is fairly invariant between regions, regardless of the specifics of the species composition. Whether this prediction is correct remains to be tested against observations.

The model was designed with fisheries applications in mind. Fishing represents a top-down perturbation of the model by an externally imposed mortality. The model may also be exposed to a bottom-up perturbation by changing the productivity or the carrying capacity of the resource. However, the response of the model to a bottom-up perturbation is less reliable than the response to a top-down perturbation because we do not know how the overall maximum recruitment is influenced by the bottom-up perturbation. The next chapter will therefore focus on the most reliable type of application, the top-down perturbation from fishing.

CHAPTER TWELVE

Community Effects of Fishing

Part II explored the impacts of fishing on demography (chapter 5), evolution (chapter 6), and dynamics (chapters 7 and 10). In this way, the theory comprehensively covered all aspects of demography needed for developing fisheries advice on a stock. By and large, many of these applications are already serviced by the classic Beverton-Holt age-based theory, with the exception of fisheries-induced evolution and consumer-resource dynamics. The size-based theory does have some advantages, such as the direct link to physiological traits for data-poor applications and the ability to estimate the recruitment. Nevertheless, the fisheries appliction developed so far falls short of the promise to deal with species interactions needed for ecosystem-based fisheries management. We did cover some aspects of species interactions by changing the physiological mortality to represent changes in the wider community leading to altered scope for growth and risk of predation, and by the consumer-resource model in chapter 10. Such simple considerations may be sufficient in a single-species advice context, however, strategic ecosystem-based management requires a more holistic approach that can directly assess how fishing on one part of the community affects other parts. For example: How does a fishery on cod affect the herring population—and vice versa? Or, more generally: How does the development of a forage fishery on small pelagic species affect the yield of a consumer fishery for large demersal species? Such ecosystem-level impact assessments are necessary to quantify the trade-offs between management actions. Trade-offs between management actions are the cornerstones in strategic management plans of entire ecosystems.

We can divide the effects of fishing into the direct effects on demography and recruitment of the targeted stocks and the indirect effects on other stocks due to the reduction in biomass of the target stocks. While the direct effects of fishing are fairly straightforward, assessing the indirect effects involves several processes. The reduction of the target stock means that competitors will face less competition—a positive impact on the rest of the fish community. However, predators on the stock will have to look for food elsewhere—a negative impact—while the stocks' prey species will enjoy a safer and more productive life—a positive impact. And it becomes even more complex: the release of the prey population from predation will increase their abundance, and they may thereby increasingly compete with juveniles from the target population—a negative impact on the target stock. Adding all these effects with different signs to assess the net outcome of the indirect effects of fishing is obviously not straightforward. The assessment is further complicated because the indirect effects propagate in all directions in the food web: individuals at the same trophic level as the target stock face decreased competition, higher trophic levels have less food, and lower trophic levels experience decreased predation. The community model developed in the previous chapter accounts for all these effects.

Modeling the impact of fishing on an entire community is more complex than the single-stock impact assessments in chapter 5. While considering a single stock in isolation, we could make a fairly exhaustive impact assessment by varying the fishing mortality, the fisheries selectivity, and the physiological mortality on all aspects of the stock: biomass, size structure, and recruitment. A similarly detailed assessment of the entire community is impossible in this space. Instead, I will focus on three important examples: trophic cascades initiated by the removal of large predators, the trade-offs between a forage fishery and a consumer fishery, and the extension of the maximum sustainable yield (MSY) concept to the community. Returning finally to the single-stock aspects, I will illustrate how the F_{msy} on each stock is context-dependent—it changes with the surrounding community.

12.1   TROPHIC CASCADES

When a component of an ecosystem is perturbed, the effects are not isolated to the component itself but cascade through the ecosystem, much like the waves from a rock thrown into a pond propagate away from the point of impact. Perturbations are mainly propagated through the predator-prey interactions: if a predator is removed, it releases the prey from predation, leading to an increase in the prey population, which then induces a higher predation pressure on the prey's prey, and so on. Such trophic cascades are the signature of indirect effects of changes in the abundance of individuals in one trophic level on other trophic levels. The previous chapter gave us an example where the high abundance of large fish led to a higher predation pressure on their prey (fig. 11.4).

A classic example of a trophic cascade is the predation by sea otters on sea urchins in the Aleutian archipelago (Estes et al., 1998). The sea otters' appetite for sea urchins kept the urchin population low, such that they were unable to graze down the kelp forest. This balance was maintained until killer whales developed a taste for sea otters. The killer whales then decimated the population of sea otters by a factor of 10 in less than a decade. The result was a huge bloom in sea urchins,

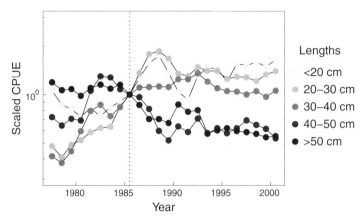

FIGURE 12.1. Catch-per-unit-effort (CPUE) from international bottom trawl survey in the North Sea organized according to length groups. All data-points are normalized with the value in 1985. Data from Daan et al. (2005).

whose increased grazing pressure led to the disappearance of kelp forests. In this example, the trophic cascade spanned four trophic levels with a strong effect all the way down to the primary producers. Similar cascades have been observed among fish communities. On the Scotian shelf, Ken Frank et al. (2005) reported the effects of disappearing groundfish stocks (largely cod) over a 20-year period. Prey species of shrimp, snow crabs, and forage fish responded to the relaxed predation pressure by increasing abundances, most pronounced among the forage fish. Increased populations of forage fish led to decreases in large zooplankton, which again resulted in higher concentrations of phytoplankton. A similar cascade was observed in the Baltic Sea, where the collapse of the cod stock led to increases in sprat populations and decreases in zooplankton (Casini et al., 2008). The observation of trophic cascades is a reminder that a perturbation on one part of the fish community—for example, the largest fish—has ecosystem-wide repercussions.

The trophic cascades can also be observed as changes in the size distribution of the community. By analyzing trawl survey data from the North Sea, Daan et al. (2005) saw that small-bodied fish increased in abundance, while large-bodied fish decreased over a 20-year period (fig. 12.1). These changes—decrease in large fish leading to a decrease in small fish—have often been described as changes in the size spectrum exponent (Rice and Gislason, 1996; Bianchi et al., 2000; Daan et al., 2005), or the *large fish indicator*, which is the ratio of biomass of individuals smaller and larger than 40 cm (Greenstreet et al., 2010; Blanchard et al., 2014). Both of these indicators can be deceptive, however, as they cannot distinguish between a trophic cascade that leads to increased numbers of small fish and one that is just a reduction in the abundance of large fish (see fig. 6 in Andersen and Pedersen, 2010, for an illustration of this effect).

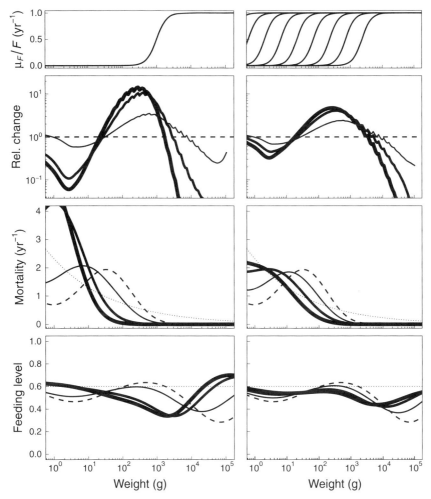

FIGURE 12.2. The impact of fishing the unexploited community from fig. 11.4 (dashed lines) with two types of fishing (top row): fishing only individuals larger then around 1 kg (left column), or fishing all species with a trawl selectivity starting around 0.05 of asymptotic size (right column). The fishing mortality is $F = 0.1, 0.3$, and $0.7 \, \text{yr}^{-1}$, shown with increasing line width. Dotted lines represent theoretical expectations, and the dashed lines are the unexploited situation. The second row shows the community size spectrum relative to the unfished spectrum, $N_c(w)/N_c(w, F = 0)$.

To explore how the modeled community responds to fishing, I expose it to two patterns of fishing in fig. 12.2: fishing on only large individuals irrespective of species (left column) and fishing on all species with a trawl-like selectivity pattern (right column). The first pattern corresponds roughly to the situation on the Scotian shelf, where the fishery is concentrated on the large demersal species, while the second pattern resembles the exploitation in a fully developed fishery where all species are full exploited, like the North Sea. In both cases, the model responds with a trophic cascade of the sort seen in the data discussed earlier: larger individuals decline, which leads to an increase in smaller fish and possibly, only in the first case, to a reduction in the zooplankton community.

The trophic cascades are first and foremost driven by changes in predation pressure: the mortality in the range 10 to 1000 g becomes smaller, while the mortality on sizes less than 10 g is increased (third row). The growth rate also has a role to play. Notice how the abundance of fish around 1000 g is increased relative to the unfished case despite being fished. The increase happens because of the higher amount of biomass entering the fished range due to the increased abundance of individuals in the size range below 1000 g. In this way, growth of individuals acts as a dampening mechanism on the trophic cascade. Therefore, the trophic cascade is damped as it moves down the trophic levels (Andersen and Pedersen, 2010). Last, changes in growth rate play a role, but it is somewhat minor because the changes are fairly small (bottom row of panels). An increase in abundance in a range leads to higher food competition, lower abundances of prey, and decreasing feeding levels and slower growth rates. In a size range where growth slows down—that is, where $df/dw < 0$, biomass piles up because it leaves a size range at a slower rate than it enters a size range. In the first fishing pattern, this occurs in the size range from about 1 to 500 g, and it therefore increases the abundance in that size range. The effect is, however, not sufficiently strong to counteract the dampening effect of the cascade by the growth between trophic levels.

12.2   WHAT IS THE IMPACT OF FORAGE FISHING?

The analysis of data and model simulations speaks clearly: the removal of large fish by fishing releases small fish from predation pressure. The resulting increase in forage fish biomass facilitates the expansion of forage fisheries. The question is, then, whether the interaction goes both ways. Will a developed forage fishery limit the productivity of large fish species (Houle et al., 2013; Ravn-Jonsen et al., 2016)? If that is the case, the developed forage fisheries will reduce the economic potential of the valuable demersal stocks or even hinder the recovery of those stocks.

To examine the interaction between forage and consumer fisheries, I have defined three fisheries: a forage fishery targeting small species (5 g $\leq W_\infty <$ 150 g),[1] a pelagic fishery targeting medium-size fish (150 g $\leq W_\infty <$ 5 kg), and a consumer fishery targeting large fish ($W_\infty \geq 5$ kg). I calculate the yield from

---

[1] I have omitted the smallest species $W_\infty = 4$ g from being exposed to forage fishing. Those small species are very vulnerable to fishing and quickly go extinct.

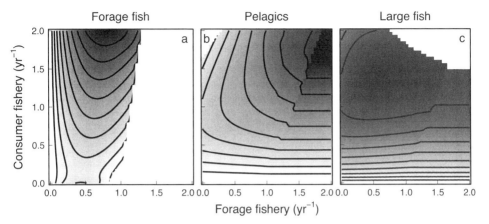

FIGURE 12.3. Yield from fishing as a function of the forage fishery (5 g $\leq W_\infty <$ 150 g) and the consumer fishery ($W_\infty \geq$ 150 g). Yield is from (a) the forage fishery; (b) small pelagic fish (150 g $\leq W_\infty <$ 5 kg); and (c) from large demersal fish ($W_\infty \geq$ 5 kg). The white area is where one asymptotic size group has been fished to extinction.

each fleet as a function of the fishing mortality on the forage fishery and on the two fisheries targeting fish for consumption (the pelagic and demersal fisheries; $W_\infty >$ 150 g) (fig. 12.3).

The yield from the forage fishery (panel a) increases as the fishing pressure in the consumer fishery is increased. This, again, illustrates how the consumer fishery facilitates the forage fishery. Turning to the yield from the fishery on large fish (panel c), we see that the contour lines are almost parallel to the x-axis, signifying that the yield is largely independent of the fishing pressure on the forage fish. Only if the fishing mortality in the consumer fishery is very high—around 1 yr^{-1}—does the model show a small negative effect of a heavy forage fishery. In other words: the forage fishery does not have a strong impact on the consumer fishery.

The limited dependence of the large fish on forage fish runs counter to our intuitive mental picture of a simple food chain: removing a basal resource—here, the forage fish—is expected to pull the rug out from under all higher trophic levels. However, the fish community cannot be perceived as a simple food chain: even fish with a large asymptotic size start their life as tiny larvae. Therefore, the removal of fish with small asymptotic size (the forage fish) releases the juvenile individuals from species with larger asymptotic size from competition. In this way, adults of the large species are compensated by the lack of forage fish prey by higher juvenile growth rates and by higher abundance of juvenile and adult medium-size fish. To which degree does this result reflect the reality of competition of predator-prey interactions in a natural ecosystem? For instance, do juveniles from large-bodied fish species share a habitat and compete for food resources with adult forage fish? Probably not to the degree assumed in model. Nevertheless, we can conclude that the propagation of trophic cascades depends upon the direction: perturbations on a high trophic level cascades down the trophic levels with a small damping for each step. Perturbations on lower trophic level species has a limited effect upon higher trophic level species.

12.3   WHAT IS THE MAXIMUM SUSTAINABLE YIELD OF A COMMUNITY?

Maximizing the yield seems like a straightforward exercise: calculate the yield as a function of fishing mortality and find the maximum—just as we did for the single stock in fig. 5.7. Things are more complicated when the entire community is considered: we need to specify how fishing mortality is distributed between species in the system.

Before going into that complication, we can explore the yield from the system when all species are fished with the same fishing mortality (fig. 12.4a). The total yield (solid line) and the biomass have a similar pattern to that in the single-species case (fig. 5.7): yield is parabolic with a well-defined maximum, and the total biomass declines (gray line). The maximum is, however, at a much higher fishing mortality than we found in chapter 5—around 2 yr^{-1} in contrast to around 0.3 yr^{-1} when a single population is exploited (fig. 5.8a). How can it be that the total community apparently can be exploited much more than a single species? The answer lies in the redistribution of predation mortality in the community, just like the trophic cascades the large-bodied species are most vulnerable to fishing, and are overexploited at fairly low fishing mortalities (as we also saw in the single-species calculations in fig. 5.8a). This releases the intermediate-size species from predation pressure so that they, despite being fished, increase in biomass by up to a factor of 5. This is the same effect we saw from the trophic cascade in fig. 12.2. The smallest species, which are also fairly vulnerable to fishing mortality (fig. 5.8a), are met with the double whammy of increased competition from the intermediate-size species combined with high fishing mortality, and consequently they decline in biomass. Eventually, at fishing mortality larger than about 0.5 yr^{-1}, the community consists almost exclusively of intermediate-size species. Even though the diversity of the community is impoverished, the yield from the fishery is high because there are no losses of these highly productive species to predation by larger species. The situation may seem academic—why would we want to overexploit the system to that degree? It has, however, been realized in practice in the East China

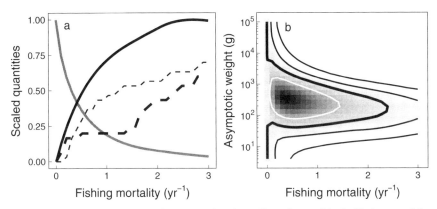

FIGURE 12.4. Yield and biomass of a community where all species are fished with the same fishing mortality. (a) Total yield (black line), total biomass (gray line), fraction of species collapsed
($F < F_{lim}$, see p. 91; thin dashed line), and fraction of species with $R_0 < 1$ (thick dashed line). (b)
Spawning stock biomass as a function of asymptotic size relative to the biomass in the unfished
situation. The thick contour line show where biomass is equal to unfished biomass; white/black
contour lines show higher/lower biomasses by factors of 2 and 5.

Sea, resulting in a fishery with a surprisingly high yield (Szuwalski et al., 2017).
The situation is reminiscent of the way agriculture is organized: by removing grazers from crops and predators from herbivores, the production is determined only
by the productivity of the basal resources (nutrients, water, and climate), and not
by losses to higher trophic levels.

How should fishing effort be distributed between species in a community in
order to achieve the highest biomass yield from the fishery (Andersen et al.,
2015)? This question confronts an omnipotent manager of all fishing fleets in an
ecosystem (fig. 12.5). The yield increases with fishing mortality, just as when
all species are evenly exploited in fig. 12.4. At high fishing mortality, the division of effort between the three fleets is fairly even. At small fishing mortalities,
however, the pattern of exploitation is different. Fishing initially targets the large-
bodied species. Only when the biomass of these species is reduced is it favorable
to develop a forage fishery. The largest species are initially targeted because they
have the highest biomass. When the biomass of the largest species is reduced, the
situation resembles the trophic cascade in fig. 12.2a: the depletion of large individuals releases smaller individuals from predation pressure. This increase in smaller
species then facilitates the development of a forage fishery. Eventually, all species
are exploited with approximately similar exploitation rates.

Ecosystem-based fisheries management is about much more than maximizing
the yield from the fishery. It is also about conservation of diversity, about securing
a high economic rent of the exploitation, and about ensuring a fair distribution of

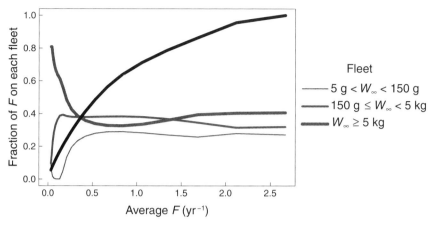

F<small>IGURE</small> 12.5. The maximum sustainable yield achieved from the community as a function of the average fishing mortality (black line). The distribution of effort between three fleets targeting small, medium, or large species are indicated with thin, medium, and thick gray lines.

the benefits throughout society. Some of these aspects can be addressed with the trait-based modeling approach. We can, for example, clearly see that striving to obtain the community maximum sustainable yield would be incompatible with the aims of ecosystem-based management because it entails that many species groups are overexploited and that the remaining community is impoverished relative to the unexploited community. Economic aspects can be addressed by modelling rent as the difference between the revenue and the costs of fishing (Gordon, 1954; Schaefer, 1954; Clark, 1973). The revenue should account for larger fish typically taking a higher price per kg than smaller ones (Andersen et al., 2015)—a kg of bluefin tuna is much more valuable than a kg of sprat. Evidently, an ecological model cannot in itself address issues of equal and fair distribution of resources in society; however, models can be used to provide the ecological baselines that are needed for socioeconomic considerations and models.

Last, we can explore how single-stock fisheries reference points are affected by the changes in the community induced by fishing (fig. 12.6). We already did this in a crude way in chapter 5 when we represented changes in the community by varying the physiological mortality (fig. 5.9). The fishing mortality leading to the maximum sustainable yield F_{msy} for each species roughly follows the same pattern as the single-stock species calculations (fig. 5.8). In general, the reference points increase as the community is fished harder. The higher productivity is facilitated by release of predation by larger species and diminished competition, both results of the lower biomass in the community. We can also see that the changes in F_{msy} are not entirely systematic; some species groups are even predicted to

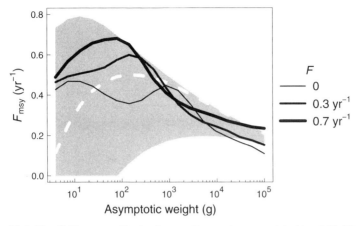

FIGURE 12.6. The fishing mortality leading to the maximum sustainable yield (F_{msy}; see section 5.3), as a function of asymptotic size. The lines show three levels of fishing on the community: $F = 0$, 0.5, and 1.0 yr^{-1}, corresponding to the communities in fig. 12.4. The white dashed line show the prediction from the single-species model from fig. 5.8, and the gray region is the range of F_{msy} if the physiological mortality constant is varied by ± 0.15 (see fig. 5.9).

tolerate lower fishing mortalities when the entire ecosystem is fished (W_∞ around 1 kg). It is therefore difficult to formulate general rules; however, the results once again highlight that fisheries reference points and the productivity of fish stocks are not constant properties of the fish stocks themselves. Fisheries reference points are context dependent—they change in tune with the dynamics of the surrounding fish community. Enlightened fisheries management therefore needs to revisit calculations of reference points continually.

12.4   SIZE- AND TRAIT-BASED MODELS FOR ECOSYSTEM-BASED FISHERIES MANAGEMENT

I have shown how the trait-based size spectrum model can be used to explore the effects of fishing on the entire ecosystem. Charles Elton (1926) wrote in his classic *Animal Ecology:* "The food-relations of animals are extremely complicated ... it is usually quite impossible to predict the precise effects of twitching one thread in the fabric." He was right, and we are still unable to accurately predict the dynamics of a specific species in a food web. However, by lumping species with similar traits—in this case, similar asymptotic sizes—the trait-based size spectrum model generates robust predictions of how fishing affects the size- and trait-structure of the fish community. This makes the model suitable as a tool for ecosystem based fisheries management.

The *ecosystem approach to fisheries management* was born out of a realization that the current approach of single-stock fisheries management is inadequate to deal with interactions between fish stock (May et al., 1979). The ecosystem approach has been formalized in international agreements (FAO, 2003) and has been adopted to various degrees and in different ways around the world. The European Union has developed the Marine Strategy Framework Directive, with the goal of "clean, healthy, and productive oceans" by 2020; the United States has tried a variety of approaches (Essington et al., 2016), and Australia has been noted for its implementation of ecosystem approaches to fisheries management, which has improved the sustainability of fished stocks (Smith et al., 2007). The ecosystem approach takes a holistic view of fisheries management by considering the entire ecosystem and not only the fish. The approach accounts for all uses and constraints of the ecosystem: production of food, generation of economic value, conservation of ecosystem function, habitats, and biodiversity, as well as equal division of ecosystem benefits among users. The goals are ambitious, and predictably management has struggled to achieve them. For one, the institutional challenges are enormous. Followed to the letter, the ecosystem approach mandates that all stakeholders—fishers, consumers, NGOs, scientists, politicians, and so on—are engaged in the process. Getting them to agree at the same table is not straightforward. The other problems are practical: we need adequate tools to make impact assessments of management actions on the ecosystem level.

Which tools does the ecosystem approach need? It is tempting to call for a replacement of single-stock models in advice with a ecosystem-oriented advice based on food-web models. However, it is clear that single-stock management will not—and should not—go away. We still need specialized working groups who know the intricacies of each specific stock, and we still need them to make stock assessments, impact assessments, and management plans. What the ecosystem approach needs is to embed these plans within a consideration of the entire ecosystem and develop *fisheries ecosystem plans,* as described by the Lenfest fishery ecosystem task force (Essington et al., 2016). Ecosystem management plans should constrain the actions of single-stock management to avoid single-stock actions that are beneficial for a specific stock but create unacceptable outcomes for other components of the ecosystem. A good example is reverse trophic cascades: when formerly overfished fish stocks recover, they impose an increasing predation pressure on forage fish. As a result, the productivity of forage fish stocks is reduced to the detriment of forage fisheries and other dependent predators such as a birds and mammals (van Gemert and Andersen, 2018a). Avoiding, or preparing, for such eventualities requires that the strategic objectives of fisheries management at the ecosystem level are explicitly formulated.

A focal point in a fisheries ecosystem plan is a vision of the ecosystem we want. From a fisheries perspective, the immediate answer might be: one that provides the maximum sustainable yield (MSY). However, MSY for the fishing fleets might compromise the needs of other dependent species, such as birds or marine mammals. Further, as we saw in this chapter, MSY in a community context is not straightforward—it is certainly different from just MSY for all the single stocks because the MSY of a stock is highly sensitive to the changes in predation pressure and, to a smaller extent, food availability (Rindorf et al., 2017). The key is that there are trade-offs between management actions, and that struggling to achieve one goal may compromise others (Link, 2010). Ecosystem models offer an appealing tool to explore some of the trade-offs between ecosystem management objectives to help develop a realistic vision for the ecosystem.

A central concern about model choice for fisheries ecosystem plans is the balance between complexity and simplicity. A complex modeling approach attempts to represent all processess in and around the ecosystem, from physics to socioeconomics, at as fine scale as possible—temporally, spatially, and for as many species as possible. An impressive example of this approach is the development of the ecosystem model system Atlantis (Fulton et al., 2011), which has been set up for ecosystems around the world. Such models provide stakeholders all the information they want and makes it possible to generate scenarios of management actions to explore the trade-offs—the costs and benefits—of management actions on the entire system. They are, however, also monstrous beasts to set up and operate. Calibration of such models for a given system requires years of effort, and running scenarios takes hours or days. What is more troublesome is their complexity. Owing to the large number of parameters and processes, many of which are poorly known, one is never sure whether all important processes are adequately resolved or whether there is a devil hidden in some detail. Their outputs of beautiful maps are beguiling to the viewer, but the interpretation of the robustness of results is difficult. Despite these problems, a well-calibrated model does generate useful output. Further, the level of detail makes them very well suited as tools to communicate with stakeholders. At the other end of the complexity-simplicity spectrum are very simple food-web motif models, such as predator-prey models or trophic chains. These models can be used for generic impact assessments (see, for example, Matsuda and Abrams, 2006). The advantages of such models are their very simple formulation with each process being clearly visible, and they are fast and easy to simulate and analyze. On the down-side, the models are poor representations of real ecosystems and they are difficult to parameterize to provide reliable quantitative estimates of rates and quantities like reference points. While such models have a place in the primary scientific literature, they are unsuited for operational fisheries management.

The size spectrum model I have developed here is sandwiched between the complex end-to-end ecosystem models and the conceptual food-web models. The size spectrum model represents only one aspect of the ecosystem—the fish community—and it does not resolve specific species but represents diversity through variation in the governing trait. However, the model is carefully derived from the individual-level processes described in chapters 2 and 3 and can produce credible quantitative estimates. In this chapter, I developed two examples of trade-offs between management actions: the initiation of a trophic cascade when large fish are overexploited, and the conflict between fisheries of forage fish and large fish. The exploration indicated some less obvious effects—in particular, that a moderate forage fishery has only a limited effect on the large species. Other explorations of ecosystem effects of fishing with the model are species recovery (Andersen and Rice, 2010), pareto efficiency (Jacobsen et al., 2017), and evaluation of *balanced harvesting* (Jacobsen et al., 2014; Kolding et al., 2016).

A good example of an attempt to make a generic fisheries ecosystem plan is the concept of balanced harvesting (Zhou et al., 2010). The idea of balanced harvesting is to distribute fishing mortality across all ecosystem components in proportion to their natural productivity in order to preserve size and species composition of the ecosystem (Garcia et al., 2012). This definition is unclear on three aspects. First, it mixes an objective—preservation of ecosystem composition—with the method—fishing proportional to productivity. Second, it is unclear what is meant by *productivity*. Productivity is generally defined as the production of a system relative to the unit of production. In fisheries, the unit of production is the fish stock, so the productivity is the production (the yield) per biomass (spawning stock biomass or fished biomass), with dimensions of time^{-1}. As we have seen repeatedly, measures related to the productivity, be it the population growth rate r_{max} or the fishing mortality at maximum sustainable yield, are highly context sensitive. Productivity is therefore not a biological property of a given stock or species, but it depends on the ecological context—in particular, the amount predators. Further, there has been confusion as how to define the unit of production, with some maintaining that the unit of production is an area, such that the productivity is the production per area with dimensions mass per time (Law et al., 2012, 2015). While this is a valid interpretation of the word *productivity*, it is a clearly a very different measure from the production per unit biomass. Third, it is not entirely evident how to define the productivity of a size class. This lack of precision in the definition has led to confusion about what exactly constitues balanced harvesting: is it a strategy that preserves ecosystem structure, or is the act of fishing proportional to productivity (whatever the definition)? Clearly, balanced harvesting in its current form does not constitute a strategic plan that can be operationalized as a fisheries ecosystem plan. It is, however, the first formulation of general principles for a plan, and as such it makes a starting line for formulating strategic fisheries ecosystem plans.

The size-spectrum model framework is not the one tool to fill all the needs of ecosystem-based fisheries management. What it can do is make simple strategic ecosystem-oriented assessments of management trade-offs, which are needed to make strategic fisheries ecosystem plans. It can also with a modest effort be calibrated more closely to resolve specific species in a given ecosystem (box 11.3), and be used for specific ecosystems (Jacobsen et al., 2017). Simulations with the model highlight how everything in the community is connected and the dynamic nature of the fisheries reference points. Fisheries management tends to treat the fisheries reference points and the productivity of stocks as fixed properties of the species. A concrete task for ecosystem based fisheries management is to provide the single-stock advice process with information about how changes in the fish community affects the mortality, and thereby the reference points, for all stocks.

PART V

EPILOGUE

CHAPTER THIRTEEN

The Size- and Trait-Based Approach

The size- and trait-based approach is a complete coherent framework for modeling fish stock and community demography and dynamics, as illustrated in fig. 1.2. The framework builds upon very few fundamental assumptions about feeding interactions (big eat small), metabolic scaling with body size, and basic principles such as energy conservation. Almost every prediction and application follows from these basic assumptions, either directly or through other derivations based on those assumptions. The assumptions gave us powerful predictions about the community size spectrum (chapter 2), the population size spectrum (chapter 4), population dynamics (chapter 7), consumer-resource dynamics (chapter 10), and the community trait structure (chapter 11). From these predictions followed applications to fisheries: reference points (chapter 5), fisheries-induced evolution (chapter 6), recovery rates (chapter 7), fisheries ecosystem plans (chapter 12), and so on, and applications to evolutionary ecology (chapters 8 and 9) and trophic cascades (chapter 11).

13.1 SIZE VERSUS AGE-BASED APPROACHES FOR FISHERIES SCIENCE

The theory is based on size as the most important characteristic of individuals. Fisheries applications, as practiced today, are predominantly performed with age-based theory. Here, I will discuss the fundamental differences between size- and age-based theory and whether using one formulation over the other brings any benefits.

The application of the theory has focused on fisheries. I have calculated fisheries reference points and the importance of size selectivity that are commonly used in contemporary fisheries management. These single-stock calculations can also be performed by classic age-based calculations as shown by Beverton and Holt (1959). So why should one use the size-based theory when another well-established framework already services those calculations? Essentially, the size-based framework is the same as the age-based; the difference lies in the variable used to structure demography (age or size). The age-based matrix calculations are based on survival probabilities from one age to the next, which were calculated in chapter 4. Note that even though the size-based approach is formulated with partial differential equations and integrals, which appear more complex than the matrices and sums in the age-based approach, the practical numerical implementation is actually with sums (box 4.4) and matrices (box 7.2), just like the age-based approach. Transforming from biomass at age to biomass at size is just a question of using the weight-at-age from the growth equation. Also, the parameters are called by different names and have different values, but I have developed transformations between them; see table A.3. There are some minor differences, though, between how current age-based theory is implemented and the size-based theory: age-based theory most often uses a constant mortality (the M), while size-based theory uses a declining mortality with size, and age-based theory uses the von Bertalanffy growth equation, while size-based theory (as formulated here) mostly uses the biphasic growth equation. These are subtle differences that do not much influence the results, and age-based calculations can easily use a size-based mortality and a biphasic growth equation. The impacts of these two adjustments on the results are for most applications modest.

The age-based approach has two advantages. First, it naturally incorporates an annual schedule. Most fish—in particular, in seasonal environments—have annual spawning cycles with either a clear spawning migration and time or a spawning period. This annual cycle is well incorporated in the age-based theory with annual time steps. Second, age-based theory is well known. The age-based approach is the de facto standard for fish stock demography. However, the age-based approach is a hindrance for the development of future applications. The size-based approach is needed to develop data-poor impact assessments (chapter 5), for data-poor stock assessments, and for the community calculations in chapter 12.

The main achievement of size-based theory is the trait-based approach. For single stocks, the trait-based approach is similar to Beverton's vision of using life-history invariants as proxies for species-specific parameters (Beverton, 1992). The trait-based approach goes beyond establishing empirical relations and roots the relations in fundamental assumptions and in the energy balance in the community. I want to stress again that the trait-based aspects does not have to be used. Everything can be described equally well at the level of specific species or populations, provided that the life-history parameters are known. When the parameters are not known, the trait-based approach offers a rough estimation of parameter values based on one or more traits. The trait-based approach revealed differences between small and large species. The most important difference is that larger species have a stronger degree of density-dependent control than small species. This insight has deep consequences for fisheries applications, as it shows why larger species are more resilient to fishing that one would expect from metabolic scaling rules, why smaller species are susceptible to environmental fluctuations, and why elasmobranchs are more sensitive to fishing than teleosts. These theoretical insights emerged clearly from the size- and trait-based approach, and would have been difficult without that.

The size-based theory for single stocks leads to important novel applications: the evolutionary and ecosystem impact assessments. These two applications are on the radar for future fisheries advice and management, but are not yet integrated. The challenge is to operationalize both applications into the advisory process of stocks and ecosystems. There are other relevant fisheres applications that I have not touched upon explicitly; two topical examples are the BOFFs and data-poor aspects.

BOFF is an acronym for big old fecund fish. The biggest females have an almost mythical appeal in the literature because of their tremendously large fecundity. Following the general scaling rules laid out in chapter 3, a newly adult mother of a species with asymptotic size 20 kg spawns about half a million eggs per year, while a BOFF at 20 kg spawns a whopping two million eggs per year (eq. 3.19 divided by an egg weight of approx. 1 mg). In general, the ratio between fecundity scales with the body weight, so the ratio between the fecundity of a newly mature female and one at the asymptotic weight is $W_\infty / w_{\mathrm{mature}} = 1/\eta_m \approx 3.5$. However, the interesting part is that some BOFFs seem also to spawn eggs and larvae with higher survival (Hixon et al., 2013), or some have a fecundity that scales faster than linear with weight (Barneche et al., 2018). These aspects increase the importance of BOFFs for reproduction. However, the BOFFs are also very rare in the population, so their importance for the reproduction of the population is diminished with respect to smaller individuals. This aspect is sometimes conveniently forgotten in the search of an attention-grabbing headline (for example, Barneche et al., 2018), but it needs to be accounted for (Berkeley et al., 2004; Farrell and Botsford, 2006; Field et al., 2008; Hixon et al., 2013; Spencer et al., 2013; Calduch-Verdiell et al., 2014). As fecundity scales with body size, the size-based approach is the obvious starting point for exploring the importance of the BOFFs, with or without maternal effects. Further, the trait-based approach offers a route to generalize such calculations to all fish life histories to single out the species with life histories (small versus large and/or fast versus slow) where the BOFFs are indeed important to consider in conservation efforts and fisheries management—and those where the BOFFs are less relevant.

Fisheries management provides stock assessment and impact assessments of many important fish stocks, but is challenged by the many "data-poor" stocks. For such stocks, not much information exists about life-history parameters and stock assessments are uncertain or absent. We lack knowledge (data) for these stocks because they are of little economic importance, or they are caught only as by-catch in other fisheries, or they are in countries unable to muster the expertise and investments in advanced management. In such cases, the size-based approach offers a way to perform stock assessment by fitting the size spectrum of catches to the population model in chapter 4, and the trait-based approach provides estimates of the relevant life-history parameters. Such approaches abound (Beddington and Kirkwood, 2005; Le Quesne and Jennings, 2012; Hordyk et al., 2014), but the trait-based approach offers also prediction of recruitment rates, which are central to estimate the reference points that are needed for managing data-poor stocks (Kokkalis et al., 2015, 2017). The applicability to data-poor situations extends to the community, where the trait-based community model can be set up to a particular situation with very little data (for example, Kolding et al., 2016), or the food-web model can be calibrated in situations where more knowledge is available (Jacobsen et al., 2017).

The size-based framework offers a route to novel models in fisheries science, provided they can be operationalized. A good example is the consumer-resource modelling framework from chapter 10. The examples I presented were idealized, as I relied on a perfectly scaling Sheldon spectrum for the resource and a perfect scaling of natural mortality, both based upon the macroecological patterns developed in chapter 2. To operationalize the consumer-resource model for a specific stock, we need better information on the size distribution of food and mortality in that specific ecosystem. Such applications have already been developed for small lake systems, where the resource is simple—small zooplankton and invertebrates where the size-structure can be ignored—and mortality is just a small background mortality or cannibalism (for example, Persson et al., 2007). It has even been extended to simple marine systems—for example, the Baltic Sea, with two main groups: forage fish and cod (van Leeuwen et al., 2008). Extending to more species rich systems is a challenge because doing so requires knowledge of ever more resources and predators. Describing the resource as a size spectrum offers a route out of the trap of increasingly complex and wobbly food-web modules. One relevant extension, though, is to consider pelagic and benthic food resource spectra—I will return to this shortly.

Applying and operationalizing consumer-resource models relates closely to the other big challenge: understanding how density dependence plays out in situ. We need to move away from blindly fitting stock-recruitment relations toward understanding the processes behind density dependence. Which exact process determines the recruitment efficiency, and which exact process determines R_{max} for a given stock? One way to move away from the stock-recruitment relation, at least partially, is to incorporate late life density dependence explicitly. Here, we must learn to disentangle interspecific and intraspecific density dependence, and distinguish between the three types of density dependence: early and late density dependence owing to competition and cannibalism. The size-based consumer-resource model offers a simulation framework to tackle these questions, and the hypothesis about spatial dynamics from section 10.3 provides a starting point for a new holistic theoretical understanding of all three types of density dependence.

13.2 FUTURE DIRECTIONS OF SIZE- AND TRAIT-BASED THEORY

An interesting open question is the nature of the trade-off between growth and mortality embodied in the physiological mortality a (see p. 157). I assumed a linear trade-off such that faster growth (higher values of A) leads to corresponding higher mortality. The linear trade-off means that the physiological mortality is the same for fast as for slow life histories (high and low values of A). However, there are indications that a varies between life histories. In the first analysis of life-history invariants, Beverton (1992) found that M/K (the adult version of a; see section 4.4) varied between different species groups, though the variance was too big to draw conclusions. A recent larger analysis by Thorson et al. (2017) found that slow-growing species like rockfish (Sebastidae) had lower values of M/K than fast-growing species like salmonids. It is then tempting to conclude that the trade-off between growth and mortality is not linear, but that faster growth results in a more than linear increase in mortality. Two things needs to be said about this conclusion. First, one should be wary of trusting the results. The statistical analysis is top-notch, but the underlying data from FishBase are not always of good quality. Mortality values are notoriously unreliable, and the estimations of growth often come with big uncertainty. Further, if the result that slower life histories have lower values of a (and M/K) than faster ones is correct, it indicates that those life histories have a higher fitness (recall that both lifetime reproductive output R_0 and the population growth rate r_{max} increase as functions of $1 - a$ (eqs. 4.39 and 7.13). That would indicate a clear evolutionary advantage of slower life histories over faster ones. Perhaps there are other trade-offs between fast and slow life histories beyond the relation between growth and mortality. A clearer view on the fast-slow trade-off requires better quality data. Assembling a high-quality data set of growth, mortality, and reproductive output for species across the range of growth rates would be a good start.

The size- and trait-based theory can serve as a framework for new applications, leading to deeper insights into the dynamics of fish communities. I here outline four exciting research questions that may be addressed with the size- and trait-based framework: stochasticity, behavior, coupling to primary production, and thermal physiology and ecology.

Stochasticity

Organisms face a highly variable environment that drives divergence in growth rates between individuals from the same population and even within the same cohort. The variability that directly affects population dynamics are differences in the encountered food; some individuals are lucky and discover abundant food, which leads to rapid early growth. Growing big fast might yield the further benefit of being able to cannibalize smaller members of the same cohort (the *lifeboat mechanism*; Gabriel, 1985; Van den Bosch et al., 1988). Differences in growth can be modeled with the size-based framework by introducing stochasticity into the feeding interaction. The effect would be to "smear out" the population size spectrum, but the main question is whether it introduces any fundamental new results or effects.

Two questions face the intrepid wanderer ino the tangled wood of stochastic population dynamics. First, one must confront the nature of the growth equation (chapter 3). The equation was formulated such that growth stopped at the asymptotic weight W_∞. However, W_∞ is not the maximum possible weight, it is the average maximum weight. In reality, some individuals, with better than average growth, will be able to grow larger, and some might terminate growth at a smaller size than W_∞. Some aspects of the differences in growth rates due to food availability is captured by the dynamic growth equation from section 3.3; however, with that equation individuals will never grow larger than W_∞ regardless of the amount of food available. To allow individuals growing larger than W_∞ requires a rethinking of the growth equation, with focus on which process exactly limits growth. It is of course possible to formulate some equation with variable maximum size, but grounding it in empirical observations requires a research program targeted at furthering our understanding of what limits growth in individuals. This cannot be done by observing size-at-age of captured wild animals, as I have done in figs. 3.3 and 9.1, but it requires repeated observations of the same individuals during ontogeny (such as in fig. A13 in Ursin, 1967). Clarifying what limits growth and the trade-offs associated with becoming larger than W_∞ might also throw new light on the BOFF issue discussed earlier. The second question is of a technical nature: How do we represent the stochasticity in the growth process— does it appear in the encounter process, or does it appear due to the finite size of each randomly encountered meal? This may seem like an esoteric question, but it will matter for the population dynamics. Stochasticity in the encounter process will lead to effects that scale with the population size, while stochasticity in the finite meal size will not (Datta et al., 2010). Whatever the answer, the technical challenges are substantial, as the complex machinery of stochastic partial differential equations is involved.

Variability may also occur in the traits. Individuals are genetically different; some individuals may be bolder, leading to faster feeding rates, while other are more timid and therefore slower growing. Such differences are tied to trade-offs, and for the bold-timid life-history axis the trade-off is that bolder individuals forage more at an increased predation risk in addition to elevated metabolic costs of activity. Such differences can be introduced as another trait-axis in the population dynamics, and the average demography can be calculated straightforwardly by summing up contributions of the different trait values. However, it becomes interesting only when individuals with different traits compete for the same resource or cannibalize one another. These questions can be addressed straightaway; however, they will probably only become really interesting when combined with the environmental stochasticity discussed here.

The exploration of the bold-timid life-history axis is of particular relevance to evolutionary fisheries management. I showed in chapter 6 how the size-selectivity of fishing gear drives selection responses of growth, size at maturation, and investment in reproduction. However, fishing gear also selects preferentially for bold or timid fish depending on whether the gear is passive or active. Bold individuals are more likely to be caught by passive gears, such as gill nets, traps, and hooks, while timid individuals appear more likely to be caught by active gears such as trawls (Arlinghaus et al., 2017). Some of the evolutionary implications have been explored (Jørgensen and Holt, 2013; Andersen et al., 2018), but not exhaustively.

Behavior

I have modeled fish as primitive organisms that follow very simple rules with regards to feeding. Essentially, they just feed as much as possible on the available food. However, fish are not that primitive; they adapt their feeding activity according to the conditions. The underlying reason is that feeding is risky: when you forage; you also risk being eaten yourself. A good example is vertical migration. It has long been known that many fish (and copepods) have a daily rhythm of vertical migration. Typically, they go up in the surface at night and down during the day (Stich and Lampert, 1981), though the reverse phenomenon is also observed (Ohman et al., 1983). It has become clear that this behavior is a response to the risk of predation: visual predators—larger fish—are particularly dangerous in the sunlit surface waters. As most of the food is also in the surface, feeding at night minimizes risk. Another good example of behavioral adaptation of feeding has been observed in Canadian lakes among juvenile charr (Biro et al., 2005). Juvenile charr can choose between two habitats: the shallow littoral zone or the open pelagic zone. The littoral zone is fairly safe because it is free from the risk of being cannibalized by adult charr, while the pelagic zone makes the juveniles susceptible to being attacked from below. Therefore, the littoral zone is a safe habitat. However, when the food in the littoral zone is grazed down, the pelagic resource becomes more attractive, and the young charr are forced to take risks. The effects of the adaptive choice of habitat on the growth rates of juvenile charr between lakes is striking: charr in high-productive and low-productive lakes have exactly the same growth rates! Clearly, this invalidates the assumption that feeding, and thus growth, is a function of the density of food. Charr in the low-productive lakes achieved similar growth rates to charr in the high-productive lakes by exposing themselves to higher risk. The difference in food conditions translated into differences in mortality, and not growth (Fiksen and Jørgensen, 2011).

The behavioral response to food and predation risk has potentially big implications for community dynamics (Jørgensen et al., 2013) because it changes a key element in the model—namely, the interaction between individuals. I have treated the interaction, represented by the feeding level, as being dependent only on the concentration of prey. With adaptive behavior, the feeding level also depends upon the concentration of predators. How does this adaptive behavior change the population and community dynamics? Does it stabilize or destabilize dynamics? How does it change the effective trophic efficiency (section 2.7)? Does it dampen or increase trophic cascades (section 12.1)?

Feeding behavior under risk can be described generally as a choice between being in a feeding arena (the upper water column or the pelagic) and a refuge (the deep or the littoral zone): being in the arena provides feeding opportunities but entails a risk, while the refuge has little food but is relatively safe. The right choice between the habitats is the one that optimizes fitness or the lifetime reproductive output. We can approximate the fitness optimization using Gilliam's rule (Gilliam and Fraser, 1987): the optimal behavior is the one that maximizes the ratio between available energy rate and mortality. In a stable environment, optimizing this ratio is the same as optimizing lifetime reproductive output, and in many situations of a variable environment it is a good approximation (Sainmont et al., 2015). Notice the similarity between Gilliam's rule and the physiological mortality a: the physiological mortality is essentially the reciprocal of Gilliam's rule divided by weight. An equivalent formulation of Gilliam's rule is therefore to minimize the physiological mortality. This observation again confirms the central role of the physiological mortality.

Let's do a quick example. Imagine a feeding arena with a food encounter rate E_e and a mortality μ, and safe refuge without food. The fraction of time spent in the feeding arena is τ. The effective food consumption is given by the functional response (eq. 10.3) to be $f = \tau E_e / (\tau E_e + C_{\max})$, and the average mortality is $\tau \mu$. The available energy is proportional to the feeding level minus respiration

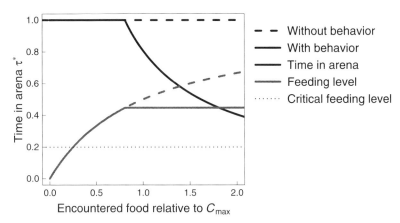

FIGURE 13.1. The average amount of time spent in the foraging arena (black lines) and the ensuing functional response (gray lines). The dashed lines show the functional response and the mortality when behavioral adaption is not accounted for.

f_c (eq. 3.30). The optimal behavior is then the one that minimizes a (eq. 4.40)

$$\tau^* = \operatorname*{argmax}_{\tau} \left\{ \frac{1}{w} \frac{\tau \mu}{\tau f - f_c} \right\} = \frac{C_{\max}}{E_e} \frac{\sqrt{f_c}}{1 - \sqrt{f_c}}. \tag{13.1}$$

If the available food $E_e < C_{\max}\sqrt{f_c}/(1+\sqrt{f_c})$, then $\tau^* > 1$. Obviously, the fish cannot spend more time than $\tau = 1$ in the feeding arena, so for low food concentrations it just spends all its time in the feeding arena. If there is more food available, it down-regulates the time in the arena with accompanying lower mortality. Fig. 13.1 summarizes these results and show the difference between the classic formulation without adaptive behavior (dashed lines) and the one with adaptive behavior (solid lines). The calculation I did here is very simplified, but it can be refined to include the predation risk in the refuge and the cost of feeding (Kiørboe et al., 2018).

The dynamic behavior in eq. 13.1 could be implemented in the consumer-resource model from chapter 10 rather straightforwardly. Implementing it in the full community model from chapter 11 is more complicated because of the multiple trophic levels in the model. In that case, one trophic level responds to its food and predators, which again elicits a response of the predators and so forth. The behavioral dynamics becomes game of multiple trophic levels (Sainmont et al., 2013). Implementing that game in a model with multiple trophic levels is a challenge; see Kondoh (2003) for an example, though with unstructured population dynamics.

Coupling to Primary Production

The exploration of the fish community in chapter 12 focused on how top-down perturbations changed the community. I showed how fishing the largest fish sets up a trophic cascade that proceeds down through the trophic levels all the way into the resource spectrum. How about the other way—how does the fish community respond to changes in primary-secondary production? Will we see a similar trophic cascade? Will a change result in new species invading and outcompeting existing ones, or will the existing species just change in abundance? How is the total productivity of the fish community affected?

One could in principle explore such bottom-up perturbations by changing the resource spectrum. Changing the carrying capacity or the productivity of the resource changes the food environment for the fish, and the model will respond by predicting changes in growth, reproduction, and fisheries productivity. Changes in the primary-secondary production would also have an impact on the carrying capacity of the stock-recruitment relation. As a start, the carrying capacity of the stock-recruitment could be changed according to the changes in the resource spectrum, following eq. 11.14. However, the stock-recruitment relation could also be changed in other ways—for example, by modifying the recruitment efficiency, ε_R. Coming to grips with the connections between the stock-recruitment relation and the primary-secondary production is central to improving the quality of bottom-up perturbations in the model.

What's more, there are different kinds of secondary production. The most important distinction is between the pelagic secondary production, mainly copepods, and the benthic secondary production, mainly invertebrates. Model-wise, this means that there are not one but two resource spectra: a pelagic and a benthic. Further, species forage to different degrees on the two resources: the pelagic resource mainly feeds pelagic specialists, such as pelagic forage fish (sprat, anchovy, sardine, and so on) and pelagic piscivores (swordfish, tuna, and so on), while the benthic resource feeds benthic specialists. The benthic and the pelagic resources are coupled by the demersal generalists that feed on both resources (Rooney et al., 2006), with cod being a good example. Pioneering efforts of resolving the lines between the two resources to the fish community have been done by Andersen and Ursin (1977) and Blanchard et al. (2011).

The productivity of the pelagic and the benthic resources are determined by the environment. Ultimately, both energy pathways are driven by the same energy source—namely, pelagic primary production. The benthic pathway is fueled by the energy lost from the pelagic production in the upper photic zone, by settling detritus or dead phytoplankton (Suess, 1980). The downward flux of matter is influenced by multiple environmental conditions. First and foremost the depth matters: in shallow shelf seas, most of the production lost from the photic zone reaches the seafloor, while in deep oceans most of the detrital matter is remineralized or turned into refractory carbon before it reaches the seafloor. Second, the proportion of primary production that is lost also varies with latitude. High-latitude regions feature a strong spring blooming of phytoplankton, mostly diatoms. These blooms are initially uncontrolled by grazing, and some of the production sinks out before predation can get them under control and retain the carbon in the photic zone (Lutz et al., 2007). Further, low water temperatures decelerate remineralization processes, so more reach the seafloor in cold high-latitude waters before it is remineralized (Pomeroy and Deibel, 1986; Laws et al., 2000). Conversely, in low latitudes, most production is retained in the upper pelagic zone. The phytoplankton production is immediately turned into secondary production, and bacterial action efficiently remineralizes detrital matter. Taken together, we expect the benthic energy pathway to be particularly strong relative to the pelagic energy pathway in shallow shelf seas in high latitudes, and these are exactly the regions where large demersal species dominates over large pelagic species (van Denderen et al., 2018). Therefore, incorporating the pelagic and energy pathways should make it possible to predict the global pattern of demersal and pelagic fish.

Thermal Physiology and Ecology

How do fish populations and communities respond to changing temperatures? As the effects of climate change on the biospehere are increasingly evident (Root et al., 2003), this question is urgent. Attention so far has focused on predicting the responses of specific populations through the direct impact of changed temperature on their physiology (for example, Pörtner and Farrell, 2008; Sinclair et al., 2016). However, increasing temperatures affect fish populations and communities on at least two time scales: On the short term is the direct physiological response to a temperature increase in terms of increasing metabolic demands. On the longer time scale is the ecological response where some species in a community will be replaced by other, better adapted, species.

The immediate physiological response to a temperature rise is an increased standard metabolic rate. Simple theoretical considerations predict that the higher energy and oxygen demands of elevated temperatures result in decreased asymptotic size (von Bertalanffy, 1957; Pörtner et al., 2017; Cheung et al., 2011). In the initial research for this book, I wanted to include a section in chapter 3 about this theory, but I dropped it because I found that it lacked empirical support. The theory has recent faced strong criticism from fish physiologists (Lefevre et al., 2017; Jutfelt et al., 2018). Resolving the impasse requires observations on individual

fish, which is very time consuming. A good starting point could be the early explorations by Ursin (1967). It will be interesting to see whether a new theoretical
understanding can be born, or whether the old theory will be propped up by new
observations. In any case, the aspect of body size will be central, and the results
can be included directly into the size-based theory by making the growth function
from chapter 3 dependent upon temperature.

The physiological response is only one aspect of how ecosystems respond to
climate change. The main action will probably be changes to the composition of
species by invasions of new species and local extinctions. Currently, the primary
tool to predict community changes is bioclimatic envelope models and species
distribution models. Such models make credible predictions of species extinctions (and possibly invasions) as the temperature in an area moves outside (or
inside) a species' thermal niche. However, populations do not occupy all sites
within their thermal niches because they are limited by the interaction with other
populations through competition, food availability, and predation. This effect is
known as the difference between the fundamental (thermal) niche and the realized niche (Hutchinson and MacArthur, 1959). Criticizing the utility of models
based on fundamental thermal niches is an old discipline. Darwin wrote (1859):
"We have reason to believe that species in a state of nature are limited in their
ranges by the competition of other organic beings quite as much as, or more than,
by adaptation to particular climates." Elton later followed up (1926): "It is fairly
useless to make elaborate 'laws of distribution' based entirely upon one factor like
temperature, as has often been done in the past. It is too crude a method." Darwin
and Elton both reject the usage of the fundamental temperature niche to predict
species distributions. Does that mean that thermal envelope models are useless
to predict the effects of climate change? Not quite. Thermal envelope models are
handy because their predictions are very robust: if the temperature moves outside
a species' thermal niche, then it is quite certain that the species will disappear
from the ecosystem. However, a problem emerges if the prediction of climate
envelope models are over interpreted. There are now many empirical and theoretical examples that show how changes in competition due to climate change are
more important than the thermal niche (Daufresne et al., 2009; Lord et al., 2017;
Zhang et al., 2017). These works show how species well within their thermal niche
are fundamentally affected by temperature changes due to their interactions with
other species. However, it is very difficult to predict such changes for specific
species. Designing models that reliably predict how competition and predation
relations change between specific species requires a heroic belief in our current
ability to understand detailed species interactions. We might be able to make
hypotheses about how specific well-studied species could respond to changing

temperatures—in particular, if said species are close to the limit of their thermal niche, but doing that for all species in a community is not possible.

Here, the trait-based approach has something to offer. If we can develop a trustworthy physiological response function of temperature change, then we can include this in the size- and trait-based community model and explore the response of the community. This exploration will not give any information about specific species, but it may provide generic predictions about changes to ecosystem functions such as production (for fisheries), trophic efficiency, and size and trait structure. Two aspects will need to be considered, though. First we need to know how interactions changes—that is, the clearance rates. Do clearance rates increase with temperature just as the metabolic demands do, or are they insensitive to temperature changes? Second, we cannot ignore changes in secondary production, discussed in the previous section. Any changes in the secondary production, both its magnitude and the division between pelagic and benthic production, will reverbate throughout the fish community. Again, the trait-based approach offers a framework to predict how. Predicting ecosystems' response to climate change is one of the biggest challenges to ecology. We cannot hope to make species-specific predictions, but predicting changes in the size- and trait-structure is an achievable goal.

I have given a comprehensive introduction to the size- and trait-based theory of fish populations and communities. The theory predicts the impact of fishing on fish stocks and communities and explains some aspects of the evolutionary ecology of fish populations. Some of those predictions will surely be challenged in the future, leading to adjustments of the underlying assumptions. Until then, the theory represents a synthesis of the current state of the art, and the predictions will serve us well as zero hypotheses. The theory holds great possibilities for further applications, and I have sketched only a few obvious areas here. I hope that the theory inspires new applications and new research to challenge the predictions, and thereby develop the theory further.

PART VI
APPENDIXES

Single Stock Size Spectrum Model

TABLE A.1. Complete Model Equations for the Size Spectrum Model
of a Single Stock with Beverton-Holt Recruitment

	Equation	References
Growth rate	$g(w) = Aw^n \left[1 - \psi_m \left(\dfrac{w}{\eta_m W_\infty} \right) \left(\dfrac{w}{W_\infty} \right)^{1-n} \right]$	Eq. 3.18
Maturation	$\psi_m(z) = [1 + z^{-5}]^{-1}$	Eq. 3.15
Predation mortality	$\mu_p(w) = aAw^{n-1}$	Eq. 4.7
Fishing mortality	$\mu_F(w) = F\psi_F(w)$	Eqs. 5.3 and 5.4
Survival	$P_{w_R \to w} = \exp\left[-\displaystyle\int_{w_R}^{w} \dfrac{\mu_p(\omega) + \mu_F(\omega)}{g(\omega)} \, d\omega \right]$	Eq. 4.9
Spectrum	$\dfrac{B(w)}{R} = \dfrac{w}{g(w)} P_{w_R \to w}$	Eq. 4.9
Spawning stock biomass	$\dfrac{B_{\mathrm{SSB}}}{R} = \displaystyle\int_{w_R}^{W_\infty} \psi_m \left(\dfrac{w}{\eta_m W_\infty} \right) \dfrac{B(w)}{R} \, dw$	
Egg production	$\dfrac{R_p}{R} = \dfrac{\varepsilon_R \varepsilon_{\mathrm{egg}}}{w_0} \left(\dfrac{w_R}{w_0} \right)^{-a} AW_\infty^{n-1} \dfrac{B_{\mathrm{SSB}}}{R}$	Eq. 4.35
Recruitment	$R/R_{\max} = 1 - (R_p/R)^{-1}$	Eq. 4.37
Fisheries yield	$Y = \displaystyle\int_{w_R}^{W_\infty} \mu_F(w) B(w) \, dw$	Eq. 5.7

Note: Box 4.4 provides the numerical solution procedure for the survival. The solutions will end up being scaled with the maximum recruitment $R_{\max}$—that is, the spectrum $B(w)/R_{\max}$, the spawning stock biomass $B_{\mathrm{SSB}}/R_{\max}$, and the recruitment $R/R_{\max}$.

TABLE A.2. Parameters Used in the Size Spectrum Model of a Single Stock

	Parameter	Value	Reference
A	Growth coefficient	$5.35\ \mathrm{g}^{0.25}\mathrm{yr}^{-1}$	Fig. 3.3
a	Physiological mortality	0.42	Fig. 4.6
η_m	Maturation relative to W_∞	0.28	Fig. 3.4
$\varepsilon_{\mathrm{egg}}$	Reproductive efficiency	0.22	Fig. 3.5
ε_R	Recruitment efficiency	0.03	†
F	Fishing mortality level	Variable	
n	Metabolic exponent	0.75	p. 24
W_∞	Asymptotic weight	Variable	
w_0	Egg weight	0.001 g	
w_R	Size at recruitment	0.001 g	

† Adjusted to obtain reference points in the right range.

TABLE A.3. Relations Between Physiological and Classic Parameters

Parameter	Relation
Asymptotic weight	$W_\infty = cL_\infty^3$
Growth coefficient	$A \approx 3c^{1/4}\eta_m^{-1/12}KL_\infty^{3/4}$
Physiological mortality	$a = \frac{1}{3}\frac{M}{K}\eta_m^{1/3}$
Recruitment efficiency	$\varepsilon_R = \alpha(\varepsilon_{\mathrm{egg}}P_{w_0 \to w_R}AW_\infty^{n-1}/w_0)^{-1}$
Maturation relative to W_∞	$\eta_m \approx (27/K^3/t_{\mathrm{mat}}^3)/64$
Asymptotic length	$L_\infty = (W_\infty/c)^{1/3}$
Von Bertalanffy growth constant	$K \approx Ac^{-1/4}\eta_m^{1/12}L_\infty^{-3/4}/3$
Adult mortality	$M = 3\eta_m^{-1/3}K$
Recruitment parameter	$\alpha = \varepsilon_R\varepsilon_{\mathrm{egg}}P_{w_0 \to w_R}AW_\infty^{n-1}/w_0$
Age at maturation	$t_{\mathrm{mat}} \approx W_\infty^{1-n}\frac{\ln(1-\eta_m^{1-n})}{A(n-1)} \approx 0.75W_\infty^{1-n}$

Consumer-Resource Model

TABLE B.1. The Consumer-Resource Model

Description	Function
Food encounter and consumption:	
Size preference	$\phi(w_p/w) = \exp\left[-(\ln(w/(\beta w_p)))^2/(2\sigma^2)\right]$
Clearance rate	$V(w) = \gamma w^q$ with $\gamma = \dfrac{f_0 h \beta^{n-q}}{(1-f_0)\sqrt{2\pi}\kappa_{res0}\sigma}$
Encountered food	$E_e(w) = V(w)\displaystyle\int_0^\infty \phi(w_p/w)(N_{res}(w) + N(w))w_p\,dw_p$
Feeding level	$f(w) = \dfrac{E_e(w)}{E_e(w) + hw^n}$
Growth and reproduction:	
Allocation to reproduction	$\psi_m(w,\,W_\infty) = \left[1 + \left(\dfrac{w}{\eta_m W_\infty}\right)^{-5}\right]^{-1}\left(\dfrac{w}{W_\infty}\right)^{1-n}$
Available energy	$E_a(w) = \begin{cases} \varepsilon_a(f(w) - f_c)hw^n & f(w) \geq f_c \\ 0 & f(w) < f_c \end{cases}$
Growth	$g(w,\,W_\infty) = E_a(w)(1 - \psi_m(w,\,W_\infty))$
Reproductive output	$R_p(W_\infty) = \dfrac{\varepsilon_R \varepsilon_{egg}}{w_0}\displaystyle\int_{w_R}^{W_\infty} N(w)E_a(w)\psi_m(w,\,W_\infty)\,dw$
Recruitment	$R(W_\infty) = R_{max}R_p/(R_p + R_{max})$
Maximum recruitment	$R_{max} = K_{Rmax}\kappa_{res0}\varepsilon_a(f_0 - f_c)hw_R^{-a}W_\infty^{2n-q-3+a}$

(Continued)

TABLE B.1. (*continued*)

Description	Function
Mortality:	
Predation (Box 10.1)	$\mu_p(w_p) = \int \phi(w_p/w)(1 - f(w))\gamma w^q N(w)\, dw$
Background mortality	$\mu_b(W_\infty) = \mu_0 w^{n-1}$
Starvation	$\mu_s(w) = \begin{cases} 0 & f(w) \geq f_c \\ -\dfrac{E_a(w)}{\xi w} & f(w) < f_c \end{cases}$
Resource dynamics:	
Resource dynamics	$\dfrac{dN_{\text{res}}(w)}{dt} = r_0 w^{n-1}\left(\kappa_{\text{res}}(w) - N_{\text{res}}(w)\right) - \mu_p(w)N_{\text{res}}(w)$
Carrying capacity	$\kappa_{\text{res}}(w) = \kappa_{\text{res}0} w^{-2-q+n}$

TABLE B.2. Parameters in the Consumer-Resource Model in Table B.1

Description	Value
Food encounter and consumption:	
Metabolic exponent	$n = 0.75$
Exponent of clearance rate	$q = 0.8$
Preferred predator-prey mass ratio	$\beta = 408$
Width of prey selection	$\sigma = 1$
Expected feeding level	$f_0 = 0.6$
Coefficient for maximum consumption	$h = 22.3 \text{ g}^{1-n}/\text{yr}$
Growth and reproduction:	
Size of maturation relative to W_∞	$\eta_m = 0.28$
Assimilation efficiency	$\varepsilon_a = 0.6$
Egg size	$w_0 = w_R = 0.001 \text{ g}$
Critical feeding level	$f_c = 0.2$
Reproduction efficiency	$\varepsilon_{\text{egg}} = 0.22$
Recruitment efficiency	$\varepsilon_R = 0.03$
Mortality:	
Background mortality	$\mu_0 = a\varepsilon_a(f_0 - f_c)h$
Physiological mortality (eq. 4.41)	$a = \beta^{2n-q-1}\dfrac{f_0}{\varepsilon_a(f_0 - f_c)}e^{(2n(q-1)-q^2+1)\sigma^2/2}$
Resource dynamics:	
Resource regeneration factor	$r_0 = 4 \text{ g}^{1-n}/\text{yr}$
Resource carrying capacity	$\kappa_{\text{res}0} = 5 \cdot 10^{-3} \text{ g}^{\lambda-1}/\text{m}^3$
Resource scaling	$\lambda = -2 - q + n$

Note: Most values are the same as used in the single-stock model from table A.2.

Community Model

TABLE C.1. Additional Equations Needed to Create the Full Trait-Based Model
From the Consumer-Resource Model in Table B.1

Description	Function
Community spectrum	$N_c(w) = N_{\text{res}}(w) + \sum\limits_{i=1}^{m} N_i(w)$
Clearance rate[1]	$V(w) = \gamma w^q$ with $\gamma = \dfrac{1.05 f_0 h \beta^{n-q}}{(1 - 1.05 f_0)\sqrt{2\pi}\kappa_{\text{res}0}\sigma}$
Background mortality	$\mu_b(W_\infty) = \mu_0 W_\infty^{n-1}$
Predation	$\mu_p(w_p) = \int \phi(w_p/w)(1 - f(w))\gamma w^q N_c(w)\,dw$
Maximum recruitment	$R_{\text{max}.i} = K_{\text{Rmax}}\kappa_{\text{res}0}\varepsilon_a(f_0 - f_c)h w_R^n W_{\infty.i}^{2n-q-3+a}\Delta W_{\infty.i}$

Note: The model is formulated with the trait-axis discretized into m asymptotic size groups. The
size spectrum of the ith group $N_i(w)$ represents all species with asymptotic sizes in the range $W_{\infty.i}$ to
$W_{\infty.i} + \Delta W_{\infty.i}$ with $\Delta W_{\infty.i} = W_{\infty.i+1} - W_{\infty.i}$. The subscript i is omitted in most equations for
clarity.

[1] γ is set such that, if the resource is at its carrying capacity, the feeding level will be $1.05 f_0$. This
procedure means that the feeding level is around f_0; see fig. 11.4b.

TABLE C.2. Parameters for the Trait-Based Model in Addition
to the Consumer-Resource Model in Table B.2

Description	Value
Asymptotic sizes	$W_\infty = 4 \ldots 10^5$ g
Width of prey selection[1]	$\sigma = 1.3$
Recruitment coefficient	$K_{\text{Rmax}} = 0.25$
Coefficient for background mortality	$\mu_0 = 1.2 \text{ g}^{1-n}/\text{yr}$
Upper size of resource spectrum	$w_{\text{cut}} = \min(W_\infty)/2 = 2$ g

[1] Increased from the usual value of 1 to represent the spread in prey preference of the different
species within an asymptotic size group.

Bibliography

Andersen, K. H. (2010). "Calculation of Expected Rates of Fisheries-Induced Evolution in Data-Poor Situations." *ICES Annual Science Conference 2010*, M:10.

Andersen, K. H., T. Berge, R. Goncalves, M. Hartvig, J. Heuschele, S. Hylander, N. S. Jacobsen, C. Lindemann, C. Martens, A. B. Neuheimer, K. Olsson, A. Palacz, F. Prowe, J. Sainmont, S. J. Traving, A. W. Visser, N. Wadhwa, and T. Kiørboe (2016). "Characteristic Sizes of Life in the Oceans, from Bacteria to Whales." *Annual Review of Marine Science* 8 : 217–241.

Andersen, K. H., and J. E. Beyer (2006). "Asymptotic Size Determines Species Abundance in the Marine Size Spectrum." *American Naturalist* 168 : 54–61.

Andersen, K. H., and J. E. Beyer (2015). "Size Structure, not Metabolic Scaling Rules, Determines Fisheries Reference Points." *Fish and Fisheries* 16 (1): 1–22.

Andersen, K. H., J. E. Beyer, and P. Lundberg (2009b). "Trophic and Individual efficiencies of size-structured communities." *Proceedings of the Royal Society of London B* 276 (1654): 109–114.

Andersen, K. H., J. E. Beyer, M. Pedersen, N. G. Andersen, and H. Gislason (2008). "Life-History Constraints on the Success of the Many Small Eggs Reproductive Strategy." *Theoretical Population Biology* 73 (4): 490–497.

Andersen, K. H., and K. Brander (2009). "Expected Rate of Fisheries-Induced Evolution Is Slow." *Proceedings of the National Academy of Science U. S. A.* 106 (28): 11657–11660.

Andersen, K. H., K. Brander, and L. Ravn-Jonsen (2015). "Trade-offs Between Objectives for Ecosystem Management of Fisheries." *Ecological Applications* 25: 1390–1396.

Andersen, K. H., K. Farnsworth, M. Pedersen, H. Gislason, and J. E. Beyer (2009a). "How Community Ecology Links Natural Mortality, Growth and Production of Fish Populations." *ICES Journal of Marine Science* 66: 1978–1984.

Andersen, K. H., N. S. Jacobsen, and K. D. Farnsworth (2016). "The Theoretical Foundations for Size Spectrum Models of Fish Communities." *Canadian Journal of Fisheries and Aquatic Science* 73 (4): 575–588.

Andersen, K. H., N. S. Jacobsen, T. Jansen, and J. E. Beyer (2017). "When in Life Does Density Dependence Occur in Fish Populations?" *Fish and Fisheries* 18: 656–667.

Andersen, K. H., L. Marty, and R. Arlinghaus (2018). "Evolution of Boldness and Life-History in Response to Selective Harvesting." *Canadian Journal of Fisheries and Aquatic Science* 75: 271–281.

Andersen, K. H., and M. Pedersen (2010). "Damped Trophic Cascades Driven by Fishing in Model Marine Ecosystems." *Proceedings of the Royal Society of London B* 277: 795–802.

Andersen, K. H., and J. C. Rice (2010). "Direct and Indirect Community Effects of Rebuilding Plans." *ICES Journal of Marine Science* 67 (9): 1980–1988.

Andersen, K. P., and E. Ursin (1977). "A Multispecies Extension to the Beverton and Holt Theory of Fishing, with Accounts of Phosphorus Circulation and Primary Production." *Meddelelser fra Danmarks Fiskeri- og Havundersøgelser* 7: 319–435.

Arlinghaus, R., K. Laskowski, J. Alos, T. Klefoth, C. T. Monk, S. Nakayama, and A. Schröder (2017). "Passive Gear-Induced Timidity Syndrome in Wild Fish Populations and Its Potential Ecological and Managerial Implications." *Fish and Fisheries* 18 (2): 360–373.

Armstrong, J. B., and D. E. Schindler (2011). "Excess Digestive Capacity in Predators Reflects a Life of Feast and Famine." *Nature* 476 (7358): 84–87.

Audzijonyte, A., A. Kuparinen, and E. A. Fulton (2013). "How Fast Is Fisheries-Induced Evolution? Quantitative Analysis of Modelling and Empirical Studies." *Evolutionary Applications* 6 (4): 585–595.

Barneche, D. R., D. R. Robertson, C. R. White, and D. J. Marshall (2018). "Fish Reproductive-Energy Output Increases Disproportionately with Body size." *Science* 360 (6389): 642–645.

Barnes, C., D. M. Bethea, R. D. Brodeur, J. Spitz, V. Ridoux, C. Pusineri, B. C. Chase, M. E. Hunsicker, F. Juanes, A. Kellermann, L. J., F. Ménard, F.-X. Bard, P. Munk, J. K. Pinnegar, F. S. Scharf, R. A. Rountree, K. I. Stergiou, C. Sassa, A. Sabates, and S. Jennings (2008). "Predator and Prey Body Sizes in Marine Food Webs." *Ecology* 89 (3): 881.

Barraquand, F. (2014). "Functional Responses and Predator-Prey Models: A Critique of Ratio Dependence." *Theoretical Ecology* 7 (1): 3–20.

Baum, J. K., R. A. Myers, D. G. Kehler, B. Worm, S. J. Harley, and P. A. Doherty (2003). "Collapse and Conservation of Shark Populations in the Northwest Atlantic." *Science* 299 (5605): 389–392.

Beddington, J. R., and G. P. Kirkwood (2005). "The Estimation of Potential Yield and Stock Status Using Life-History Parameters." *Proceedings of the Royal Society B* 360 (1453): 163–170.

Benoît, E., and M.-J. Rochet (2004). "A Continuous Model of Biomass Size Spectra Governed by Predation and the Effects of Fishing on Them." *Journal of Theoretical Biology* 226 (1): 9–21.

Benoît, H. P., D. P. Swain, W. D. Bowen, G. A. Breed, M. O. Hammill, and V. Harvey (2011). "Evaluating the Potential for Grey Seal Predation to Explain Elevated Natural Mortality in Three Fish Species in the Southern Gulf of St. Lawrence." *Marine Ecology Progress Series* 442: 149–167.

Berkeley, S. A., M. A. Hixon, R. J. Larson, and M. S. Love (2004). "Fisheries Sustainability via Protection of Age Structure and Spatial Distribution of Fish Populations." *Fisheries* 29 (8): 23–32.

Beverton, R.J.H. (1992). "Patterns of Reproductive Strategy Parameters in Some Marine Teleost Fishes." *Journal of Fish Biology* 41: 137–160.

Beverton, R.J.H., and S. J. Holt (1957). *On the Dynamics of Exploited Fish Populations.* Fishery Investigation Series II(19). London: Her Majesty's Stationary office.

Beverton, R.J.H., and S. J. Holt (1959). "A Review of the Lifespans and Mortality Rates of Fish in Nature and the Relation to Growth and Other Physiological Characteristics." In *Ciba Foundation Colloquia in Aging. V. The Lifespan of Animals*, pp. 142–177. London: Churchill.

Beyer, J. E. (1989). "Recruitment Stability and Survival—Simple Size-Specific Theory with Examples from the Early Life Dynamic of Marine Fish." *Dana* 7: 45–147.

Bianchi, G., H. Gislason, K. Graham, L. Hill, X. Jin, K. Koranteng, S. Manickchand-Heileman, I. Paya, K. Sainsbury, F. Sanchez et al. (2000). "Impact of Fishing on Size Composition and Diversity of Demersal Fish Communities." *ICES Journal of Marine Science* 57 (3): 558–571.

Biro, P. A., J. R. Post, and M. V. Abrahams (2005). "Ontogeny of Energy Allocation Reveals Selective Pressure Promoting Risk-Taking Behaviour in Young Fish Cohorts." *Proceedings of the Royal Society of London B: Biological Sciences* 272 (1571): 1443–1448.

Blanchard, J. L., K. H. Andersen, F. Scott, N. T. Hintzen, G. Piet, and S. Jennings (2014). "Evaluating Targets and Trade-offs among Fisheries and Conservation Objectives Using a Multispecies Size Spectrum Model." *Journal of Applied Ecology* 51 (3): 612–622.

Blanchard, J. L., R. Law, M. D. Castle, and S. Jennings (2011). "Coupled Energy Pathways and the Resilience of Size-Structured Food Webs." *Theoretical Ecology* 4 (3): 289–300.

Borgmann, U. (1987). "Models on the Slope of, and Biomass Flow up, the Biomass Size Spectrum." *Canadian Journal of Fisheries and Aquatic Science* 44 (Supp. 2): 136–140.

Borrell, B. (2013). "A Big Fight over Little Fish." *Nature* 493 (7434): 597.

Boudreau, P. R., and L. M. Dickie (1992). "Biomass Spectra of Aquatic Ecosystems in Relation to Fisheries Yield." *Canadian Journal of Fisheries and Aquatic Science* 49 (8): 1528–1538.

Brander, K. M. (1981). "Disappearance of Common Skate *Raia batis* from Irish Sea." *Nature* 290: 48–49.

Brown, J. H., J. F. Gillooly, A. P. Allen, V. M. Savage, and G. B. West (2004). "Toward a Metabolic Theory of Ecology." *Ecology* 85 (7): 1771–1789.

Brundtland, G., M. Khalid, S. Agnelli, S. Al-Athel, B. Chidzero, L. Fadika, V. Hauff, I. Lang, M. Shijun, M. Morino de Botero, M. Singh, S. Okita et al. (1987). *Our Common Future ("Brundtland Report")*. Oxford, UK: Oxford University Press.

Burrough, R., and C. Kennedy (1979). "The Occurrence and Natural Alleviation of Stunting in a Population of Roach *Rutilus rutilus* (L.)." *Journal of Fish Biology* 15(1): 93–109.

Calduch-Verdiell, N., K. H. Andersen, L. Ravn-Jonsen, B. R. Mackenzie, and J. W. Vaupel (2011). "Ecological and Economic Consequences of Different Recovery Scenarios of Depleted Stocks." *ICES Annual Science Conference 2011*, M:03.

Calduch-Verdiell, N., B. R. MacKenzie, J. W. Vaupel, and K. H. Andersen (2014). "A Life-History Evaluation of the Impact of Maternal Effects on Recruitment and Fisheries Reference Points." *Canadian Journal of Fisheries and Aquatic Sciences* 71(7): 1113–1120.

Carpenter, S. R., J. F. Kitchell, J. R. Hodgson, P. A. Cochran, J. J. Elser, M. M. Elser, D. M. Lodge, D. Kretchmer, X. He, and C. N. von Ende (1987). "Regulation of Lake Primary Productivity by Food Web Structure." *Ecology* 68(6): 1863–1876.

Casini, M., J. Lövgren, J. Hjelm, M. Cardinale, J.-C. Molinero, and G. Kornilovs (2008). "Multi-level Trophic Cascades in a Heavily Exploited Open Marine Ecosystem." *Proceedings of the Royal Society of London B* 275(1644): 1793–1801.

Charnov, E. L., and J. F. Gillooly (2004). "Size and Temperature in the Evolution of Fish Life Histories." *Integrative and Comparative Biology* 44(6): 494–497.

Charnov, E. L., H. Gislason, and J. G. Pope (2013). "Evolutionary Assembly Rules for Fish Life Histories." *Fish and Fisheries* 14(2): 213–224.

Charnov, E. L., T. F. Turner, and K. O. Winemiller (2001). "Reproductive Constraints and the Evolution of Life Histories with Indeterminate Growth." *Proceedings of the National Academy of Science U. S. A.* 98(16): 9460–9464.

Cheung, W. W., J. Dunne, J. L. Sarmiento, and D. Pauly (2011). "Integrating Ecophysiology and Plankton Dynamics into Projected Maximum Fisheries Catch Potential under Climate Change in the Northeast Atlantic." *ICES Journal of Marine Science* 68(6): 1008–1018.

China, V., and R. Holzman (2014). "Hydrodynamic Starvation in First-Feeding Larval Fishes." *Proceedings of the National Academy of Sciences of the United States of America* 111(22): 8083–8088.

Clark, C. W. (1973). "The Economics of Overexploitation." *Science* 181(4100): 630–634.

Conover, D. O., and S. B. Munch (2002). "Sustaining Fisheries Yields over Evolutionary Time Scales." *Science* 297: 94–96.

Conover, D. O., S. B. Munch, and S. A. Arnott (2009). "Reversal of Evolutionary Downsizing Caused by Selective Harvest of Large Fish." *Proceedings of the Royal Society B* 276(1664): 2015–2020.

Daan, N., H. Gislason, J. G. Pope, and J. C. Rice (2005). "Changes in the North Sea Fish Community: Evidence of Indirect Effects of Fishing?" *ICES Journal of Marine Science* 62(2): 177–188.

Damuth, J. (1987). "Interspecific Allometry of Population Density in Mammals and Other Animals: The Independence of Body Mass and Population Energy-Use." *Biological Journal of the Linnean Society* 31(3): 193–246.

Darwin, C. (1859). *On the Origin of Species by Means of Natural Selection*. London: J. Murray.

Datta, S., G. W. Delius, and R. Law (2010). "A Jump-Growth Model for Predator–Prey Dynamics: Derivation and Application to Marine Ecosystems." *Bulletin of Mathematical Biology* 72(6): 1361–1382.

Daufresne, M., K. Lengfellner, and U. Sommer (2009). "Global Warming Benefits the Small in Aquatic Ecosystems." *Proceedings of the National Academy of Sciences of the U. S. A.* 106(31): 12788–12793.

De Roos, A. M. (1988). "Numerical Methods for Structured Population Models: The Escalator Boxcar Train." *Numerical Methods for Partial Differential Equations* 4(3): 173–195.

De Roos, A. M., and L. Persson (2002). "Size-Dependent Life-History Traits Promote Catastrophic Collapses of Top Predators." *Proceedings of the National Academy of Sciences of the U. S. A.* 99(20): 12907–12912.

De Roos, A. M., and L. Persson (2003). "Competition in Size-Structured Populations: Mechanisms Inducing Cohort Formation and Population Cycles." *Theoretical Population Biology* 63(1): 1–16.

De Roos, A. M., and L. Persson (2013). *Population and Community Ecology of Ontogenetic Development*. Princeton, NJ: Princeton University Press.

De Roos, A. M., T. Schellekens, T. van Kooten, K. van de Wolfshaar, D. Claessen, and L. Persson (2007). "Food-Dependent Growth Leads to Overcompensation in Stage-Specific Biomass When Mortality Increases: The Influence of Maturation versus Reproduction Regulation." *American Naturalist* 170(3): E59–76.

De Ruiter, P. C., A. M. Neutel, and J. C. Moore (1995). "Energetics, Patterns of Interaction Strengths, and Stability in Real Ecosystems." *Science* 269(5228): 1257–1260.

Diaz Pauli, B., M. Wiech, M. Heino, and A. C. Utne-Palm (2015). "Opposite Selection on Behavioural Types by Active and Passive Fishing Gears in a Simulated Guppy *Poecilia reticulata* Fishery." *Journal of Fish Biology* 86(3): 1030–1045.

Dingsør, G. E., L. Ciannelli, K.-S. Chan, G. Ottersen, and N. C. Stenseth (2007). "Density Dependence and Density Independence During the Early Life Stages of Four Marine Fish Stocks." *Ecology* 88(3): 625–634.

Dunlop, E. S., M. Heino, and U. Dieckmann (2009). "Eco-Genetic Modeling of Contemporary Life-History Evolution." *Ecological Applications* 19(7): 1815–1834.

Easterling, M. R., S. P. Ellner, and P. M. Dixon (2000). "Size-Specific Sensitivity: Applying a New Structured Population Model." *Ecology* 81(3): 694–708.

Economo, E. P., A. J. Kerkhoff, and B. J. Enquist (2005). "Allometric Growth, Life-History Invariants and Population Energetics." *Ecology Letters* 8: 353–360.

Edwards, A. M., J.P.W. Robinson, M. J. Plank, J. K. Baum, and J. L. Blanchard (2017). "Testing and Recommending Methods for Fitting Size Spectra to Data." *Methods in Ecology and Evolution* 8(1): 57–67.

Eero, M., M. Vinther, H. Haslob, B. Huwer, M. Casini, M. Storr-Paulsen, and F. W. Köster (2012). "Spatial Management of Marine Resources Can Enhance the Recovery of Predators and Avoid Local Depletion of Forage Fish." *Conservation Letters* 5(6): 486–492.

Eikeset, A. M., E. S. Dunlop, M. Heino, G. Storvik, N. C. Stenseth, and U. Dieckmann (2016). "Roles of Density-Dependent Growth and Life History Evolution in Accounting for Fisheries-Induced Trait Changes." *Proceedings of the National Academy of Sciences of the U. S. A.* 113(52): 15030–15035.

Elton, C. S. (1926). *Animal Ecology*. Chicago: University of Chicago Press.

Essington, T. E., J. F. Kitchell, and C. J. Walters (2001). "The von Bertalanffy Growth Function, Bioenergetics, and the Consumption Rates of Fish." *Canadian Journal of Fisheries and Aquatic Sciences* 58(11): 2129–2138.

Essington, T. E., P. S. Levin, K. N. Marshall, L. E. Koehn, L. G. Anderson, A. Bundy, C. Carothers, F. Coleman, J. H. Grabowski, L. R. Gerber, L. R. Houde, E. O. Jensen, C. Mãllmann, K. Rose, J. N. Sanchirico, and A.D.M. Smith (2016). "Building Effective Fishery Ecosystem Plans: A Report from the Lenfest Fishery Ecosystem Task Force." Technical Report, Washington, DC: Lenfest Ocean Program.

Estes, J. A., M. T. Tinker, T. M. Williams, and D. F. Doak (1998). "Killer Whale Predation on Sea Otters Linking Oceanic and Nearshore Ecosystems." *Science* 282(5388): 473–476.

Evans, G. T. and J. S. Parslow (1985). "A Model of Annual Plankton Cycles." *Biological Oceanography* 3(3): 327–347.

Falster, D. S., A. T. Moles, and M. Westoby (2008). "A General Model for the Scaling of Offspring Size and Adult Size." *The American Naturalist* 172(3): 299–317.

FAO (2003). "Fisheries Management 2. The Ecosystem Approach to Fisheries." Technical Report, Food and Agriculture Organization, Rome.

FAO (2016). "The State of World Fisheries and Aquaculture." Technical Report, Food and Agriculture Organization, Rome.

Farrell, M.R.O., and L. W. Botsford (2006). "The Fisheries Management Implications of Maternal-Age-Dependent Larval Survival." *Canadian Journal of Fisheries and Aquatic Sciences* 2258: 2249–2258.

Fenchel, T. (1974). "Intrinsic Rate of Natural Increase: The Relationship with Body Size." *Oecologia* 14: 317–326.

Field, J. G., C. L. Moloney, L. du Buisson, A. Jarre, T. Stroemme, M. R. Lipinski, and P. Kainge (2008). "Exploring the BOFFFF Hypothesis Using a Model of Southern African Deepwater Hake (*Merluccius paradoxus*)." In *Fisheries for Global Welfare and Environment, 5th World Fisheries Congress*, pp. 17–26. Tokyo: Terrapub.

Fiksen, Ø., and C. Jørgensen (2011). "Model of Optimal Behaviour in Fish Larvae Predicts That Food Availability Determines Survival, but Not Growth." *Marine Ecology Progress Series* 432: 207–219.

Ford, J. R., and S. E. Swearer (2013). "Two's Company, Three's a Crowd: Food and Shelter Limitation Outweigh the Benefits of Group Living in a Shoaling Fish." *Ecology* 94(5): 1069–1077.

Fox, L. R. (1975). "Cannibalism in Natural Populations." *Annual Review of Ecology and Systematics* 6(1): 87–106.

Frank, K. T., B. Petrie, J. S. Choi, and W. C. Leggett (2005). "Trophic Cascades in a Formerly Cod-Dominated Ecosystem." *Science* 308(5728): 1621–1623.

Froese, R. (2006). "Cube Law, Condition Factor and Weight-Length Relationships: History, Meta-analysis and Recommendations." *Journal of Applied Ichthyology* 22(4): 241–253.

Froese, R., and D. Pauly (2017). *FishBase*. World Wide Web electronic publication. Available at www.fishbase.org.

Fulton, E. A., J. S. Link, I. C. Kaplan, M. Savina-Rolland, P. Johnson, C. Ainsworth, P. Horne, R. Gorton, R. J. Gamble, A.D.M. Smith, and D. C. Smith (2011). "Lessons in Modelling and Management of Marine Ecosystems: The Atlantis Experience." *Fish and Fisheries* 12(2): 171–188.

Gabriel, W. (1985). *Overcoming Food Limitation by Cannibalism*. Ed. W. Lampert. Proceedings of Internation Symposium, Piön, Germany, July 9–13, 1984.

Garcia, S. M., J. Kolding, J. Rice, M. J. Rochet, S. Zhou, T. Arimoto, J. E. Beyer, L. Borges, A. Bundy, D. Dunn et al. (2012). "Reconsidering the Consequences of Selective Fisheries." *Science* 335(6072): 1045–1047.

Gilliam, J. F., and D. F. Fraser (1987). "Habitat Selection under Predation Hazard: Test of a Model with Foraging Minnows." *Ecology* 68(6): 1856–1862.

Gislason, H., N. Daan, J. C. Rice, and J. G. Pope (2010). "Size, Growth, Temperature and the Natural Mortality of Marine Fish." *Fish and Fisheries* 11(2): 149–158.

Goodwin, N. B., A. Grant, A. L. Perry, N. K. Dulvy, and J. D. Reynolds (2006). "Life History Correlates of Density-Dependent Recruitment in Marine Fishes." *Canadian Journal of Fisheries and Aquatic Science* 63(3): 494–509.

Gordon, H. (1954). "The Economic Theory of a Common Property Resource: The Fishery." *Journal of Political Economy* 62: 124–142.

Grafton, R. Q., T. Kompas, L. Chu, and N. Che (2010). "Maximum Economic Yield." *Australian Journal of Agricultural and Resource Economics* 54(3): 273–280.

Graham, M. (1948). "Reporter's Review of the Scientific Meeting on the Effect of the War on the Stocks of Commercial Food Fishes." In *Rapports et Proces-Verbaux des Reunions du Conseil International pour l'Exploration de la Mer*, 122: 6.

Greenstreet, S.P.R., S. I. Rogers, J. C. Rice, G. J. Piet, E. J. Guirey, H. M. Fraser, and R. J. Fryer (2010). "Development of the EcoQO for the North Sea Fish Community." *ICES Journal of Marine Science* 68(1): 1–11.

Grime, J. (1977). "Evidence for the Existence of Three Primary Strategies in Plants and Its Relevance to Ecological and Evolutionary Theory." *American Naturalist* 111(982): 1169–1194.

Gunderson, D. R. (1997). "Trade-off between Reproductive Effort and Adult Survival in Oviparous and Viviparous Fishes." *Canadian Journal of Fisheries and Aquatic Science* 54(5): 990–998.

Haldane, J. (1928). "On Being the Right Size." In *A Treasury of Science* ed. H. Shapely, S. Raffort, and H. Wright, pp. 321–325. New York: Harper.

Hartvig, M. (2011). "Ecological Processes Yield Complex and Realistic Food Webs." In *Food Web Ecology*, Ph. D. thesis, pp. 73–100. Lund University.

Hartvig, M., and K. H. Andersen (2013). "Coexistence of Structured Populations with Size-Based Prey Selection." *Theoretical Population Biology* 89: 24–33.

Hartvig, M., K. H. Andersen, and J. E. Beyer (2011). "Food Web Framework for Size-Structured Populations." *Journal of Theoretical Biology* 272(1): 113–122.

Hilborn, R. (2010). "Pretty Good Yield and Exploited Fishes." *Marine Policy* 34: 193–196.

Hilborn, R., and K. Stokes (2010). Defining Overfished Stocks: Have We Lost the Plot? *Fisheries* 35(3): 113–120.

Hilborn, R., and C. J. Walters (1992). *Quantitative Fisheries Stock Assessment: Choice, Dynamics and Uncertainty*. Dordrecht: Science & Business Models.

Hinrichsen, H.-H., M. St. John, E. Aro, P. Grønkjær, and R. Voss (2001). "Testing the Larval Drift Hypothesis in the Baltic Sea: Retention versus Dispersion caused by Wind-Driven Circulation." *ICES Journal of Marine Science* 58(5): 973–984.

Hirst, A. G., and T. Kiørboe (2002). "Mortality of Marine Planktonic Copepods: Global Rates and Patterns." *Marine Ecology Progress Series* 230: 195–209.

Hixon, M. A., D. W. Johnson, and S. M. Sogard (2013). "BOFFFFs: On the Importance of Conserving Old-Growth Age Structure in Fishery Populations." *ICES Journal of Marine Science* 71(8): 2171–2185.

Hixon, M. A., and G. P. Jones (2005). "Competition, Predation, and Density-Dependent Mortality in Demersal Marine Fishes." *Ecology* 86(11): 2847–2859.

Hjort, J. (1914). "Fluctuations in the Great Fisheries of Northern Europe." *Rapports et Proces-verbaux des Réunions du Conseil International pour l'Exploration de la Mer* 20: 1–228.

Holden, M. (1973). "Are Long-Term Sustainable Fisheries for Elasmobranchs Possible?" *Rapports et Proces-verbaux des Réunions du Conseil International pour l'Exploration de la Mer* 164: 360–367.

Holt, S. (2006). "The Notion of Sustainability." In *Gaining Ground: In Pursuit of Ecological Sustainability*, ed. D. M. Lavigue, pp. 43–82 Guelph, Canada: International Fund for Animal Welfare, and Limerick, Ireland, University of Limerick.

Hordyk, A., K. Ono, S. Valencia, N. Loneragan, and J. Prince (2014). "A Novel Length-Based Empirical Estimation Method of Spawning Potential Ratio (SPR), and Tests of Its Performance, for Small-scale, Data-Poor Fisheries." *ICES Journal of Marine Science* 72(1): 217–231.

Houle, J. E., K. H. Andersen, K. D. Farnsworth, and D. G. Reid (2013). "Emerging Asymmetric Interactions Between Forage and Predator Fisheries Impose Management Trade-offs." *Journal of Fish Biology* 83(4): 890–904.

Hutchings, J. A., R. A. Myers, V. B. García, L. O. Lucifora, and A. Kuparinen (2012). "Life-History Correlates of Extinction Risk and Recovery Potential." *Ecological Applications* 22(4): 1061–1067.

Hutchinson, G. E., and R. H. MacArthur (1959). "A Theoretical Ecological Model of Size Distributions among Species of Animals." *American Naturalist* 93(869): 117–125.

ICES (2000). "Report of the CWP Intersessional Meeting Working Group on Precautionary Approach Terminology and CWP Sub-group on Publication of Integrated Catch Statistics for the Atlantic." Technical Report, ICES. ICES CM 2000/ACFM:17.

ICES (2018). "Report of the Working Group on Ecosystem Effects of Fishing Activities (WGECO)." Technical Report, ICES. ICES CM 2018/ACOM:27.

Jacobsen, N. S., M. G. Burgess, and K. H. Andersen (2017). "Efficiency of Fisheries is Increasing at the Ecosystem Level." *Fish and Fisheries* 18(2): 199–211.

Jacobsen, N. S., H. Gislason, and K. H. Andersen (2014). "The Consequences of Balanced Harvesting of Fish Communities." *Proceedings of the Royal Society B* 281(1775): 20132701.

Jennings, S. (2007). "Measurement of Body Size and Abundance in Tests of Macroecological and Food Web Theory." *Journal of Animal Ecology* 44(1): 72–82.

Jennings, S., F. Mélin, J. L. Blanchard, R. M. Forster, N. K. Dulvy, and R. W. Wilson (2008). "Global-Scale Predictions of Community and Ecosystem Properties from Simple Ecological Theory." *Proceedings of the Royal Society of London B* 275: 1375–1383.

Jørgensen, C., K. Enberg, E. S. Dunlop, R. Arlinghaus, D. S. Boukal, K. Brander, B. Ernande, A. Gårdmark, F. Johnston, S. Matsumura et al. (2007). "Managing Evolving Fish Stocks." *Science* 318(5854): 1247–1248.

Jørgensen, C., B. Ernande, and Ø. Fiksen (2009). "Size-Selective Fishing Gear and Life History Evolution in the Northeast Arctic Cod." *Evolutionary Applications* 2(3): 356–370.

Jørgensen, C., and Ø. Fiksen (2006). "State-Dependent Energy Allocation in Cod *Gadus morhua*." *Canadian Journal of Fisheries and Aquatic Science* 63: 186–199.

Jørgensen, C., and R. E. Holt (2013). "Natural Mortality: Its Ecology, How It Shapes Fish Life Histories, and Why It May Be Increased by Fishing." *Journal of Sea Research* 75: 8–18.

Jørgensen, C., A. F. Opdal, and Ø. Fiksen (2013). "Can Behavioural Ecology Unite Hypotheses for Fish Recruitment?" *ICES Journal of Marine Science* 71(4): 909–917.

Jutfelt, F., T. Norin, R. Ern, J. Overgaard, T. Wang, D. J. McKenzie, S. Lefevre, G. E. Nilsson, N. B. Metcalfe, A. J. Hickey et al. (2018). "Oxygen- and Capacity-Limited Thermal Tolerance: Blurring Ecology and Pphysiology." *Journal of Experimental Biology* 221(1): jeb169615.

Kerr, S. R. (1974). "Theory of Size Distribution in Ecological Communities." *Journal Fisheries Research Board of Canada* 31(12): 1859–1862.

Killen, S. S., I. Costa, J. A. Brown, and A. K. Gamprel (2007). "Little Left in the Tank: Metabolic Scaling in Marine Teleosts and Its Implications for Aerobic Scope." *Proceedings of the Royal Society of London B* 274: 431–438.

Killen, S. S., D. S. Glazier, E. L. Rezende, T. D. Clark, D. Atkinson, A.S.T. Willener, and L. G. Halsey (2016). "Ecological Influences and Morphological Correlates of Resting and Maximal Metabolic Rates across Teleost Fish Species." *American Naturalist* 187(5): 592–606.

Killen, S. S., J.J.H. Nati, and C. D. Suski (2015). "Vulnerability of Individual Fish to Capture by Trawling Is Influenced by Capacity for Anaerobic Metabolism. *Proceedings of the Royal Society B* 282(1813): 20150603.

Kiørboe, T. (2011). "What Makes Pelagic Copepods So Successful?" *Journal of Plankton Research* 33(5): 677–685.

Kiørboe, T., and A. G. Hirst (2014). "Shifts in Mass Scaling of Respiration, Feeding, and Growth Rates across Life-Form Transitions in Marine Pelagic Organisms." *American Naturalist* 183(4): E118–30.

Kiørboe, T., E. Saiz, P. Tiselius, and K. H. Andersen (2018). "Adaptive Feeding Behavior and Functional Responses in Zooplankton." *Limnology and Oceanography* 63(1): 308–321.

Kiørboe, T., A. Visser, and K. H. Andersen (2018). "A Trait-Based Approach to Ocean Ecology." *ICES Journal of Marine Science*, 75(6) 1849–1863. doi:10.1093 /icesjms/fsy090.

Kitchell, J. F., D. J. Stewart, and D. Weininger (1977). "Applications of a Bioenergetics Model to Yellow Perch (*Perca flavescens*) and Walleye (*Stizostedion vitreum vitreum*)." *Journal Fisheries Research Board of Canada* 34: 1922–1935.

Kleiber, M. (1932). "Body Size and Metabolism." *Hilgardia* 6: 315–353.

Kokkalis, A., A. M. Eikeset, U. H. Thygesen, P. Steingrund, and K. H. Andersen (2017). "Estimating Uncertainty of Data Limited Stock assessments." *ICES Journal of Marine Science* 74(1): 69–77.

Kokkalis, A., U. H. Thygesen, A. Nielsen, and K. H. Andersen (2015). "Reliability of Fisheries Reference Points Estimation for Data-Poor Stocks." *Fisheries Research* 107: 4–11.

Kolding, J., N. S. Jacobsen, K. H. Andersen, and P. van Zwieten (2016). "Maximizing Fisheries Yields while Maintaining Community Structure." *Canadian Journal of Fisheries and Aquatic Science* 73(4): 644–655.

Kondoh, M. (2003). "Foraging Adaptation and the Relationship Between Food-Web Complexity and Stability." *Science* 299: 1388–1391.

Kooijman, S.A.L.M. (2000). *Dynamic Energy and Mass Budgets in Biological Systems.* Cambridge, UK: Cambridge University Press.

Kurlansky, M. (1998). *Cod: A Biography of the Fish That Changed the World.* London: Penguin Books.

Lankford, T., J. Billerbeck, and D. Conover (2001). "Evolution of Intrinsic Growth and Energy Acquisition Rates. II. Trade-offs with Vulnerability to Predation in *Menidia menidia.*" *Evolution* 55(9): 1873–1881.

Larkin, P. (1977). "An Epitaph for the Concept of Maximum Sustained Yield." *Transactions of the American Fisheries Association* 106(1): 1–11.

Laugen, A. T., G. H. Engelhard, R. Whitlock, R. Arlinghaus, D. J. Dankel, E. S. Dunlop, A. M. Eikeset, K. Enberg, C. Jørgensen, S. Matsumura, S. Nusslé, D. Urbach, L. C.

Baulier, D. S. Boukal, B. Ernande, F. D. Johnston, F. Mollet, H. Pardoe, N. O. Therk-ildsen, S. Uusi-Heikkilä, A. Vainikka, M. Heino, A. D. Rijnsdorp, and U. Dieckmann (2014). "Evolutionary Impact Assessment: Accounting for Evolutionary Consequences of Fishing in an Ecosystem Approach to Fisheries Management." *Fish and Fisheries* 15(1): 65–96.

Law, R. (2000). "Fishing, Selection, and Phenotypic Evolution." *ICES Journal of Marine Science* 57(3): 659–668.

Law, R. (2007). "Fisheries-Induced Evolution: Present Status and Future Directions." *Marine Ecology Progress Series* 335: 271–277.

Law, R., and D. R. Grey (1989). "Evolution of Yields from Populations with Age-Specific Cropping." *Evolutionary Ecology* 3: 343–359.

Law, R., J. Kolding, and M. J. Plank (2015). "Squaring the Circle: Reconciling Fishing and Conservation of Aquatic Ecosystems." *Fish and Fisheries* 16(1): 160–174.

Law, R., M. J. Plank, and J. Kolding (2012). "Marine Science from Dynamic Size Spectra." *ICES Journal of Marine Science* 69(2011): 602–614.

Law, R., M. J. Plank, and J. Kolding (2016). "Balanced Exploitation and Coexistence of Interacting, Size-Structured, Fish Species." *Fish and Fisheries* 17(2): 281–302.

Laws, E. A., P. G. Falkowski, W. O. Smith Jr., H. Ducklow, and J. J. McCarthy (2000). "Temperature Effects on Export Production in the Open Ocean." *Global Biogeochemical Cycles* 14(4): 1231–1246.

Lefevre, S., D. J. McKenzie, and G. E. Nilsson (2017). "Models Projecting the Fate of Fish Populations under Climate Change Need to Be Based on Valid Physiological Mechanisms." *Global Change Biology* 23(9): 3449–3459.

Le Quesne, W. J., and S. Jennings (2012). "Predicting Species Vulnerability with Minimal Data to Support Rapid Risk Assessment of Fishing Impacts on Biodiversity." *Journal of Applied Ecology* 49(1): 20–28.

Lester, N. P., B. J. Shuter, and P. A. Abrams (2004). "Interpreting the von Bertalanffy Model of Somatic Growth in Fishes: The Cost of Reproduction." *Proceedings of the Royal Society of London B* 271: 1625–1631.

Lewy, P., and M. Vinther (2004). "A Stochastic Age-Length-Structured Multi-Species Model Applied to North Sea Stocks." *ICES Annual Science Conference 2004*, FF:20.

Lindeman, R. L. (1942). "The Trophic Aspect of Ecology." *Ecology* 23: 399–418.

Link, J. (2010). *Ecosystem-Based Fisheries Management: Confronting Tradeoffs.* Cambridge, UK: Cambridge University Press.

Lord, J., J. Barry, and D. Graves (2017). "Impact of Climate Change on Direct and Indirect Species Interactions." *Marine Ecology Progress Series* 571: 1–11.

Lorenzen, K., and K. Enberg (2002). "Density-Dependent Growth as a Key Mechanism in the Regulation of Fish Populations: Evidence from Among-Population Comparisons." *Proceedings of the Royal Society of London B* 269(1486): 49–54.

Lutz, M. J., K. Caldeira, R. B. Dunbar, and M. J. Behrenfeld (2007). "Seasonal Rhythms of Net Primary Production and Particulate Organic Carbon Flux to Depth Describe the Efficiency of Biological Pump in the Global Ocean." *Journal of Geophysical Research: Oceans* 112(C10).

MacArthur, R., and R. Levins (1967). "The Limiting Similarity, Convergence, and Divergence of Coexisting Species." *American Naturalist* 101(921): 377–385.

Martens, E. A., N. Wadhwa, N. S. Jacobsen, C. Lindemann, K. H. Andersen, and A. Visser (2015). "Size Structures Sensory Hierarchy in Ocean Life." *Proceedings of the Royal Society B* 282: 20151346.

Matsuda, H., and P. A. Abrams (2006). "Maximal Yields from Multispecies Fisheries Systems: Rules for Systems with Multiple Trophic Levels." *Ecological Applications* 16(1): 225–237.

Matsumura, S., R. Arlinghaus, and U. Dieckmann (2011). "Assessing Evolutionary Consequences of Size-Selective Recreational Fishing on Multiple Life-History Traits, with an Application to Northern Pike (*Esox lucius*)." *Evolutionary Ecology* 25(3): 711–735.

Maury, O., and J. C. Poggiale (2013). "From Individuals to Populations to Communities: A Dynamic Energy Budget Model of Marine Ecosystem Size-Spectrum Including Life History Diversity." *Journal of Theoretical Biology* 324: 52–71.

May, R. M., J. R. Beddington, C. W. Clark, S. J. Holt, and R. M. Laws (1979). "Management of Multispecies Fisheries." *Science* 205(4403): 267–277.

McGurk, M. D. (1986). "Natural Mortality of Marine Pelagic Fish Eggs and Larvae: Role of Spatial Patchiness." *Marine Ecology Progress Series* 34: 227–242.

Metz, J.A.J., and O. Diekmann (1986). *The Dynamics of Physiologically Structured Populations*, Vol. 68. Berlin: Springer.

Mollet, F., U. Dieckmann, and A. Rijnsdorp (2016). "Reconstructing the Effects of Fishing on Life History Evolution in North Sea Plaice (*Pleuronectes platessa*)." *Marine Ecology Progress Series* 542: 195–208.

Munch, S. B., M. L. Snover, G. M. Watters, and M. Mangel (2005). "A Unified Treatment of Top-Down and Bottom-Up Control of Reproduction in Populations." *Ecology Letters* 8(7): 691–695.

Munk, W., and G. Riley (1952). "Absorption of Nutrients by Aquatic Plants." *Journal of Marine Research* 11: 215–240.

Murdoch, W. W. (1969). "Switching in General Predators: Experiments on Predator Specificity and Stability of Prey Populations." *Ecological Monographs* 39(4): 335–354.

Murdoch, W. W., S. Avery, and M. E. Smyth (1975). "Switching in Predatory Fish." *Ecology* 56(5): 1094–1105.

Myers, R. A., and N. Cadigan (1993). "Density-Dependent Juvenile Mortality in Marine Demersal Fish." *Canadian Journal of Fisheries and Aquatic Science* 50(8): 1576–1590.

Myers, R. A. (2001). "Stock and Recruitment: Generalizations about Maximum Reproductive Rate, Density Dependence, and Variability Using Metaanalytic Approaches." *ICES Journal of Marine Science* 58(5): 937–951.

Myers, R. A., N. J. Barrowman, J. A. Hutchings, and A. A. Rosenberg (1995). "Population Dynamics of Exploited Fish Stocks at Low Population Levels." *Science* 269(5227): 1106–1108.

Myers, R. A., and J. M. Hoenig (1997). "Direct Estimates of Gear Selectivity from Multiple Tagging Experiments." *Canadian Journal of Fisheries and Aquatic Sciences* 54(1): 1–9.

Mylius, S. D., and O. Diekmann (1995). "On Evolutionarily Stable Life Histories, Optimization and the Need to Be Specific about Density Dependence." *Oikos* 74(2): 218–224.

Neuheimer, A. B., M. Hartvig, J. Heuschele, S. Hylander, T. Kiørboe, K. Olsson, J. Sain-mont, and K. H. Andersen (2015). "Adult and Offspring Size in the Ocean over 17 Orders of Magnitude Follows Two Life-History Strategies." *Ecology* 96(12): 3303–3311.

Ohman, M. D., B. W. Frost, and E. B. Cohen (1983). "Reverse Diel Vertical Migration: An Escape from Invertebrate Predators." *Science* 220(4604): 1404–1407.

Olsen, E. M., M. Heino, G. R. Lilly, M. J. Morgan, J. Brattey, B. Ernande, and U. Dieck-mann (2004). "Maturation Trends Indicative of Rapid Evolution Preceded the Collapse of Northern Cod." *Nature* 428: 932–935.

Olsson, K., and H. Gislason (2016). "Testing Reproductive Allometry in Fish." *ICES Journal of Marine Science* 73(6): 1466–1473.

Olsson, K., H. Gislason, and K. H. Andersen (2016). "Differences in Density-Dependence Drive Dual Offspring Size Strategies in Fish." *Journal Theoretical Biology* 407: 118–127.

Persson, L., P.-A. Amundsen, A. M. de Roos, A. Klementsen, R. Knudsen, and R. Prim-icerio (2007). "Culling Prey Promotes Predator Recovery—Alternative States in a Whole-Lake Experiment." *Science* 316: 1743–1746.

Persson, L., K. Leonardsson, A. M. de Roos, B. Gyllenberg, and M. Christensen (1998). "Ontogenetic Scaling of Foraging Rates and the Dynamics of a Size-Structured Consumer-Resource Model." *Theoretical Population Biology* 54: 270–293.

Persson, L., A. van Leeuwen, and A. M. de Roos (2014). "The Ecological Foundation for Ecosystem-Based Management of Fisheries: Mechanistic Linkages Between the Individual-, Population-, and Community-Level Dynamics." *ICES Journal of Marine Science* 71(8): 2268–2280.

Petchey, O. L., A. P. Beckerman, J. O. Riede, and P. H. Warren (2008). "Size, Foraging, and Food Web Structure." *Proceedings of the National Academy of Sciences of the U. S. A.* 105(11): 4191–4196.

Pianka, E. R. (1970). "On *r*- and *K*-selection." *American Naturalist* 104(940): 592–597.

Pimm, S. L., J. H. Lawton, and J. E. Cohen (1991). "Food Web Patterns and Their Consequences." *Nature* 350(6320): 669–674.

Pomeroy, L. R., and D. Deibel (1986). "Temperature Regulation of Bacterial Activity During the Spring Bloom in Newfoundland Coastal Waters." *Science* 233(4761): 359–361.

Pope, J. G., J. C. Rice, N. Daan, S. Jennings, and H. Gislason (2006). "Modelling an Exploited Marine Fish Community with 15 Parameters—Results from a Simple Size-Based Model." *ICES Journal of Marine Science* 63(6): 1029–1044.

Pörtner, H. O., and A. Farrell (2008). "Physiology and Climate Change." *Science* 322 (October): 690–692.

Pörtner, H.-O., C. Bock, and F. C. Mark (2017). "Oxygen- and Capacity-Limited Thermal Tolerance: Bridging Ecology and Physiology." *Journal of Experimental Biology* 220(15): 2685–2696.

Press, W. (2007). *Numerical Recipes, 3rd Edition: The Art of Scientific Computing.* Cambridge, UK: Cambridge University Press.

Punt, A. E., D. C. Smith, and A.D.M. Smith (2011). "Among-Stock Comparisons for Improving Stock Assessments of Data-Poor Stocks: The 'Robin Hood' Approach." *ICES Journal of Marine Science* 68(5): 972–981.

Quince, C., P. A. Abrams, B. J. Shuter, and N. P. Lester (2008). "Biphasic Growth in Fish I: Theoretical Foundations." *Journal of Theoretical Biology* 254(2): 197–206.

Quinn, T. J., and J. S. Collie (2005). "Sustainability in Single-Species Population Models." *Philosophical Transactions of the Royal Society of London Series B* 360(1453): 147–162.

Quinn, T. J., and R. Deriso (1999). *Quantitative Fish Dynamics*. Oxford, UK: Oxford University Press.

Rask, M. (1983). "Differences in Growth of Perch (*Perca fluviatilis* L.) in Two Small Forest Lakes." *Hydrobiologia* 101: 139–144.

Ravn-Jonsen, L., K. H. Andersen, and N. Vestergaard (2016). "An Indicator for Ecosystem Externalities in Fishing." *Natural Resource Modeling* 29(3): 400–425.

Reuman, D. C., C. Mulder, D. Raffaelli, and J. E. Cohen (2008). "Three Allometric Relations of Population Density to Body Mass: Theoretical Integration and Empirical Tests in 149 Food Webs." *Ecology Letters* 11(11): 1216–1228.

Rice, J., and H. Gislason (1996). "Patterns of Change in the Size Spectra of Numbers and Diversity of the North Sea Fish Assemblages, as Reflected in Surveys and Models." *ICES Journal of Marine Science* 53: 1214–1225.

Rijnsdorp, A. D., and P. V. Leeuwen (1992). "Density-Dependent and Independent Changes in Somatic Growth of Female North Sea Plaice *Pleuronectes platessa* Between 1930 and 1985 as Revealed by Back-Calculation of Otoliths." *Marine Ecology Progress Series* 88: 19–32.

Rijnsdorp, A. D., and P. V. Leeuwen (1996). "Changes in Growth of North Sea Plaice since 1950 in Relation to Density, Eutrophication, Beam-Trawl Effort, and Temperature." *ICES Journal of Marine Science* 53(6): 1199–1213.

Rijnsdorp, A. D. (1993). "Fisheries as a Large-Scale Experiment on Life-History Evolution—Disentangling Phenotypic and Genetic-Effects in Changes in Maturation and Reproduction of North-Sea Plaice, *Pleuronectes platessa* L." *Oecologia* 96: 391–401.

Rindorf, A., C. M. Dichmont, P. S. Levin, P. Mace, S. Pascoe, R. Prellezo, A. E. Punt, D. G. Reid, R. Stephenson, C. Ulrich, M. Vinther, and L. W. Clausen (2017). "Food for Thought: Pretty Good Multispecies Yield." *ICES Journal of Marine Science* 74: 475–486.

Ripple, W. J., C. Wolf, T. M. Newsome, M. Hoffmann, A. J. Wirsing, and D. J. McCauley (2017). "Extinction Risk Is Most Acute for the World's Largest and Smallest Vertebrates." *Proceedings of the National Academy of Sciences* 114(40): 10678–10683.

Rooney, N., K. McCann, G. Gellnet, and J. C. Moore (2006). "Structural Asymmetry and the Stability of Diverse Food Webs." *Nature* 442(20): 265–269.

Root, T., J. Price, K. Hall, and S. Schneider (2003). "Fingerprints of Global Warming on Wild Animals and Plants." *Nature* 421(6918): 57–60.

Rosenzweig, M. L., and R. H. MacArthur (1963). "Graphical Representation and Stability Conditions of Predator-Prey Interactions." *American Naturalist* 97(895): 209–223.

Rossberg, A. (2013). *Food Webs and Biodiversity: Foundations, Models, Data*. West Sussex, UK: Wiley.

Sainmont, J., K. H. Andersen, U. H. Thygesen, Ø. Fiksen, and A. W. Visser (2015). "An Effective Algorithm for Approximating Adaptive Behavior in Seasonal Environments." *Ecological Modelling* 311: 20–30.

Sainmont, J., U. H. Thygesen, and A. W. Visser (2013). "Diel Vertical Migration Arising in a Habitat Selection Game." *Theoretical Ecology* 6(2): 241–251.

Schaefer, M. B. (1954). "Some Aspects of the Dynamics of Populations Important to the Management of Commercial Marine Fisheries." *Bulletin of Inter-American Tropical Tuna Commission* 1(2): 25–56.

Sheldon, R. W., A. Prakash, and W. H. Sutcliffe (1972). "The Size Distribution of Particles in the Ocean." *Limnology and Oceanography* 17(3): 327–340.

Sheldon, R. W., and S. R. Kerr (1972). "The Population Density of Monsters in Loch Ness." *Limnology and Oceanography* 17(5): 796–798.

Sheldon, R. W., and T. R. Parsons (1967). "A Continuous Size Spectrum for Particulate Matter in the Sea." *Journal Fisheris Research Board of Canada* 24(5): 909–915.

Sheldon, R. W., W. H. Sutcliffe Jr., and M. A. Paranjape (1977). "Structure of Pelagic Food Chain and Relationship Between Plankton and Fish Production." *Journal Fisheries Research Board of Canada* 34: 2344–2353.

Shepherd, J. G. (1982). "A Versatile New Stock-Recruitment Relationship for Fisheries, and the Construction of Sustainable Yield Curves." *Journal Conseil Internatineux Exploration du Mer* 40(1): 67–75.

Silvert, W., and T. Platt (1978). "Energy Flux in the Pelagic Ecosystem: A Time-Dependent Equation." *Limnology and Oceanography* 23(4): 813–816.

Sinclair, B. J., K. E. Marshall, M. A. Sewell, D. L. Levesque, C. S. Willett, S. Slotsbo, Y. Dong, C. D. Harley, D. J. Marshall, B. S. Helmuth et al. (2016). "Can We Predict Ectotherm Responses to Climate Change Using Thermal Performance Curves and Body Temperatures?" *Ecology Letters* 19(11): 1372–1385.

Smith, A., E. Fulton, A. Hobday, D. Smith, and P. Shoulder (2007). "Scientific Tools to Support the Practical Implementation of Ecosystem-Based Fisheries Management." *ICES Journal of Marine Science* 64(4): 633–639.

Smith, C., and P. Reay (1991). "Cannibalism in Teleost Fish." *Reviews in Fish Biology and Fisheries* 1: 41–64.

Spence, M. A., P. Blackwell, and J. Blanchard (2015). "Parameter Uncertainty of a Dynamic Multi-Species Size Spectrum Model." *Canadian Journal of Fisheries and Aquatic Science* 73(4): 589–597.

Spencer, P. D., S. B. Kraak, and E. A. Trippel (2013). "The Influence of Maternal Effects in Larval Survival on Fishery Harvest Reference Points for Two Life-History Patterns." *Canadian journal of fisheries and aquatic sciences* 71(1): 151–161.

Sprules, W., and L. Barth (2016). "Surfing the Biomass Size Spectrum: Some Remarks on History, Theory, and Application." *Canadian Journal of Fisheries and Aquatic Sciences* 73(4): 477–495.

Stevens, J. D., R. Bonfil, N. K. Dulvy, and P. A. Walker (2000). "The Effects of Fishing on Sharks, Rays, and Chimaeras (Chondrichthyans), and the Implications for Marine Ecosystems." *ICES Journal of Marine Science* 57: 476–494.

Stevenson, C., L. S. Katz, F. Micheli, B. Block, K. W. Heiman, C. Perle, K. Weng, R. Dunbar, and J. Witting (2007). "High Apex Predator Biomass on Remote Pacific Islands." *Coral Reefs* 26(1): 47–51.

Stich, H.-B., and W. Lampert (1981). "Predator Evasion as an Explanation of Diurnal Vertical Migration by Zooplankton." *Nature* 293(5831): 396–399.

Suess, E. (1980). "Particulate Organic Carbon Flux in the Oceans-Surface Productivity and Oxygen Utilization." *Nature* 288(5788): 260–263.

Szuwalski, C. S., M. G. Burgess, C. Costello, and S. D. Gaines (2017). "High Fishery Catches Through Trophic Cascades in China." *Proceedings of the National Academics of Science U. S. A.* 114: 717–721.

Thorson, J. T., S. B. Munch, J. M. Cope, and J. Gao (2017). "Predicting Life History Parameters for All Fishes Worldwide." *Ecological Applications* 27(8): 2262–2276.

Thygesen, U. H., K. Farnsworth, K. H. Andersen, and J. E. Beyer (2005). "How Optimal Life History Changes with the Community Size-Spectrum." *Proceedings of the Royal Society of London B* 272(1570): 1323–1331.

Ursin, E. (1967). "A Mathematical Model of Some Aspects of Fish Growth, Respiration and mortality." *Journal Fisheries Research Board Canada* 24(11): 2355–2453.

Ursin, E. (1973). "On the Prey Size Preferences of Cod and Dab." *Meddelelser fra Danmarks Fiskeri- og Havundersøgelser* 7: 85–98.

Ursin, E. (1979). "Principles of Growth in Fishes." *Symposic of the Zoological Society of London* 44: 63–87.

Van den Bosch, F., A. De Roos, and W. Gabriel (1988). "Cannibalism as a Life Boat Mechanism." *Journal of Mathematical Biology* 26(6): 619–633.

van Denderen, P. D., M. Lindegren, B. R. MacKenzie, R. A. Watson, and K. H. Andersen (2018). "Global Patterns in Marine Predatory Fish." *Nature Ecology and Evolution* 2(1): 65–69.

van der Veer, H. (1986). "Immigration, Settlement, and Density-Dependent Mortality of a Larval and Early Postlarval 0-Group Plaice (*Pleuronectes platessa*) Population in the Western Wadden Sea." *Marine Ecology Progress Series* 29: 223–236.

van Gemert, R., and K. H. Andersen (2018a). "Challenges to Fisheries Advice and Management Due to Stock Recovery." *ICES Journal of Marine Science* 75(4): 1296–1305.

van Gemert, R., and K. H. Andersen (2018b). "Implications of Late-in-Life Density-Dependent Growth for Fishery Size-at-Entry Leading to Maximum Sustainable Yield." *ICES Journal of Marine Science* 75(4): 1296–1305.

van Leeuwen, A., A. de Roos, and L. Persson (2008). "How Cod Shapes Its World." *Journal of Sea Research* 60(1–2): 89–104.

von Bertalanffy, L. (1957). "Quantitative Laws in Metabolism and Growth." *Quarterly Review of Biology* 32(3): 217–231.

Walters, C., D. Pauly, V. Christensen, and J. F. Kitchell (2000). "Representing Density Dependent Consequences of Life History Strategies in Aquatic Ecosystems: Ecosim II." *Ecosystems* 3(1): 70–83.

Wang, Z. J. (2000). "Two Dimensional Mechanism for Insect Hovering." *Physical Review Letters* 85(10): 2216

Ware, D. M. (1978). "Bioenergetics of Pelagic Fish: Theoretical Change in Swimming Speed and Ration with Body Size." *Journal Fisheries Research Board of Canada* 35: 220–228.

Werner, E. E., and J. F. Gilliam (1984). "The Ontogenetic Niche and Species Interactions in Size-Structured Populations." *Annual Review of Ecology and Systematics* 15: 393–425.

West, G. B., J. H. Brown, and B. J. Enquist (1997). "A General Model for the Origin of Allometric Scaling Laws in Biology." *Science* 276(5309): 122–126.

West, G. B., J. H. Brown, and B. J. Enquist (2001). "A General Model for Ontogenetic Growth." *Nature* 413: 628–631.

Westoby, M., D. S. Falster, A. T. Moles, P. A. Vesk, and I. J. Wright (2002). "Plant Ecological Strategies: Some Leading Dimensions of Variation Between Species." *Annual Review of Ecology and Systematics* 33(1): 125–159.

White, E. P., S. K. M. Ernest, A. J. Kerkhoff, and B. J. Enquist (2007). "Relationships Between Body Size and Abundance in Ecology." *Trends in Ecology and Evolution* 22(6): 323–330.

Winemiller, K. O., and K. A. Rose (1992). "Patterns of Life-History Diversification in North American Fishes: Implications for Population Regulation." *Canadian Journal of Fisheries and Aquatic Sciences* 49(10): 2196–2218.

Yletyinen, J., W. Butler, G. Ottersen, K. H. Andersen, S. Bonanomi, F. K. Diekert, C. Folke, M. Lindegren, M. C. Nordström, A. Richter, L. Rogers, G. Romagnoni, B. Weigel, J. D. Whittington, T. Blenckner, and N. C. Stenseth (2018). "When Is a Fish Stock Collapsed?" *bioRxiv*: 329979.

Ylikarjula, J., M. Heino, and U. Dieckmann (1999). "Ecology and Adaptation of Stunted Growth in Fish." *Evolutionary Ecology* 13: 433–453.

Yodzis, P., and S. Innes (1992). "Body Size and Consumer-Resource Dynamics." *American Naturalist* 139(6): 1151–1175.

Zanden, M.J.V., and W. W. Fetzer (2007). "Global Patterns of Aquatic Food Chain Length." *Oikos* 116(8): 1378–1388.

Zhang, L., M. Hartvig, M. Knudsen, and K. H. Andersen (2014). "Size-Based Predictions of Food Web Patterns." *Theoretical Ecology* 7: 23-330.

Zhang, L., K. H. Andersen, U. Dieckmann, and Å. Brännström (2015). "Four Types of Interference Competition and Their Impacts on the Ecology and Evolution of Size-Structured Populations and Communities." *Journal of Theoretical Biology* 380: 280–290.

Zhang, L., D. Takahashi, M. Hartvig, and K. H. Andersen (2017). "Food-Web Dynamics under Climate Change." *Proceedings of the Royal Society B* 284(1867): 20171772.

Zhou, S., A.D.M. Smith, A. E. Punt, A. J. Richardson, M. Gibbs, E. A. Fulton, S. Pascoe, C. Bulman, P. Bayliss, and K. Sainsbury (2010). "Ecosystem-Based Fisheries Management Requires a Change to the Selective Fishing Philosophy." *Proceedings of the National Academy of Science U. S. A.* 107(21): 9485–9489.

Zhou, S., S. Yin, J. T. Thorson, and A.D.M. Smith (2012). "Linking Fishing Mortality Reference Points to Life History Traits: An Empirical Study." *Canadian Journal of Fisheries and Aquatic Sciences* 69(8): 1292–1301.

Zijlema, M. (1996). "On the Construction of a Third-Order Accurate Monotone Convection Scheme with Application to Turbulent Flows in General Domains." *International Journal for Numerical Methods in Fluids* 22: 619–641.

Zimmermann, F., D. Ricard, and M. Heino (2018). "Density Regulation in Northeast Atlantic Fish Populations: Density Dependence Is Stronger in Recruitment Than in Somatic Growth." *Journal of Animal Ecology* 87(3): 672–681.

Index

MONOGRAPHS IN POPULATION BIOLOGY

SIMON A. LEVIN AND HENRY S. HORN, SERIES EDITORS